Sustainability

The Institute for Policy Analysis and Development was established to promote research into and the analysis of the various factors of human existence and to develop policy options for a transition to a sustainable way of life, including but without prejudice to the generality of the foregoing:

To promote research into and the analysis of the philosophical, psychological, social, economic, political, technological, environmental and other factors that determine the parameters for human existence, thereby to identify those elements, trends, or patterns that are leading to or are likely to lead to significant environmental or social degradation or destruction and consequent suffering and damaging disruption of human social and economic systems, and to develop policy options for a transition to a sustainable way of life that does not cause significant environmental or social degradation or destruction, does not significantly compromise the lives of future generations, and utilises the resources of the planet wisely to ensure a relatively indefinite future for human society and other forms of life on Earth.

To publish and promote the findings of this research and the conclusions and results thereof, and to engage in related or other research, educational, promotional, or advisory activities.

The Institute for Policy Analysis and Development
60 Cumberland Street
Edinburgh
EH3 6RE
Telephone/Fax: (01578) 760248
International: +44 1578 760248

Sustainability

A SYSTEMS APPROACH

Anthony M H Clayton
Nicholas J Radcliffe

WestviewPress
A Division of HarperCollins*Publishers*

Published in 1996 in the United Kingdom by
Earthscan Publications Limited
120 Pentonville Road
London, N1 9JN

Published in 1996 in the United States of America by
Westview Press, Inc
5500 Central Avenue
Boulder
Colorado 80301-2877

ISBN 0-8133-3184-6 (hc) 0-8133-3185-4 (pb)
Library of Congress Cataloguing-in-Publication Data available on request.

Contents

Foreword

The human species is facing an extraordinary challenge. We must make decisions that no species has ever had to contemplate before. We must consciously determine the ecology of our own species.

Life on Earth has evolved into a unique, complex and beautiful phenomenon, in which there is both change and stability. The stability results from interlocking checks and balances, in which every species plays its role with little or no awareness of the true complexity of the ecological, biological, biochemical, geochemical, and physical dynamics that constitute the system of which it forms a part. The rate and scale of human impact on the global ecology is such that it has now become necessary to think about these system dynamics, and whether it is possible that our species could engineer its own decline or even demise.

Much of the current impact on the global environment is caused by the rapid expansion of human activity into a range of environments. This is happening with little conscious strategy, and without much reckoning of potential consequences. Some part of the impact is being caused to little or no real advantage.

The fundamental challenge of sustainability goes far beyond that of environmentalism. The question is whether we can fulfil our unique potential as human beings, to understand our behaviour and its consequences. To do this, we must be prepared to discard our prejudices, and to review every area of human life. We must transcend the current limitations on our thinking if we are to become aware and rational beings in a way that no other species has ever had to do or been able to do before.

Whether we have sufficient vision and resolve to do this remains to be seen.

Acknowledgements

The report on which this book is based was the result of a study commissioned by WWF Scotland in 1990. The members of staff who commissioned and managed the study were Simon Pepper, Dominic White and Elizabeth Leighton. The original report, and this book, were produced by the Institute for Policy Analysis and Development. The principal author of the report was Dr Anthony M H Clayton, the secondary author was Professor Nicholas J Radcliffe, while advisory and other support was provided by Ray Ross, Richard S Smith, Dr Monir Tayeb and Ian Thomson of the Institute.

A number of other people contributed to the original report. Simon Pepper and Martin Mathers, Dominic White, and Elizabeth Leighton of WWF Scotland not only initiated the study, provided funding and reviewed intermediate drafts, but also provided essential guidance and other support at a number of critical stages in the work, while Heather Henderson and Alison Colquhoun provided administrative and secretarial support in, respectively, early and late phases of the study.

Especial thanks for their expert advice and guidance are due to the members of the interim review panel, Robin Callander, Dr Roddy Fairley of Scottish Natural Heritage, and Professor Rob Gray of the University of Dundee; to those who reviewed early drafts, including Sandy Cumming of Highlands and Islands Enterprise, Vincent Goodstadt of Strathclyde Regional Council, Peter Ramshaw of WWF-UK, and Clive Spash of the University of Stirling; to those who reviewed later drafts, including Stephen Bass of the International Institute for Environment and Development, Robin Callander, Craig Campbell of the Scottish Council of Development and Industry, Dr Roddy Fairley of Scottish Natural Heritage, Professor Rob Gray of the University of Dundee, Nicholas Gubbins of Scottish Natural Heritage, Tom Hart of the University of Glasgow, Professor Bill Heal of the Institute of Terrestrial Ecology, Jenny Heap of English Nature, Dr Peter Hopkinson and John Bowers of the Universities of Bradford and Leeds, Roger Levett of CAG Consultants, Dr Ulrich Loening of the Centre for Human Ecology at the University of Edinburgh, Rick Minter of the Countryside Commission, Dr Ian Moffat of the

University of Stirling, Alan Mowle of the Scottish Office Environment Department, Dr Philip Ratcliffe of the Forestry Authority, Richard Sandbrook of the International Institute for Environment and Development, Professor John Smythe of the Scottish Environmental Education Council, Diane Walters of the Heriot-Watt University and David Westwood of UK2000 Scotland; and to those who provided additional advice in the final revision, including Professor Rob Gray and Professor John Smythe.

The final form and content of this book and any errors are the responsibility of the authors.

Anthony M H Clayton and Nicholas J Radcliffe

During the course of the study on which this book is based I became seriously ill. The support I received at that time, from my family and friends, helped to ensure my own survival. The debt I owe to my friends and colleagues, including those in the Institute and in WWF Scotland, is not just an intellectual debt.

Tony Clayton

Preface

The question of Sustainability affects all areas of human activity. The range of literature relevant to the subject is now extremely large, and many definitions of sustainability are now in use – some incompatible. There are a number of generalist publications that develop relevant proposals for more participative teaching methods and planning processes, the introduction of new economic indicators that incorporate some environmental costs, reform of the procedures for funding development in the less developed nations and so on. Many of these are extremely pertinent, but they are disparate and often fail to connect to an underlying analysis that could link these suggestions together into a coherent rationale and programme for change. Other publications give relevant practical suggestions in the form of guidelines for greening your business or your local authority, insulating your house, changing your lifestyle and so on. Many of these publications are useful, but they usually contain relatively little information on the principles, theory and research behind the techniques and recommendations, or about the political and economic impediments to more fundamental change. Then there is the more specialist literature in both natural and social science. These scientific papers often discuss important methodological and theoretical issues, which are vital to the stimulation and development of scholarly research, but are often inaccessible to those who are not highly numerate, or who are unfamiliar with the area or who do not have a a scientific background.

This book has been written partly in order to fill the middle ground between the more applied publications and the more specialist books and papers. It will not be able to fill all needs. It has been necessary to make many decisions, some very difficult, as to what to include. The treatment is selective rather than exhaustive. Although many would agree that a multi-disciplinary approach is necessary to understand the extensive and complex ramifications of sustainability, specialists often resist comment from outside their discipline, arguing that non-specialists cannot fully understand the complexity of the issues. We have touched on some of what we believe are the important links between a number of areas, and thereby risk offending a great many specialists. We should make it clear, therefore, that our reviews of these fields are quite selective and informed by our particular perspective. We do not attempt to give a comprehensive review of all of the relevant fields, as this purpose is served by the existing specialist literature.

The most important purpose of this book, however, is to describe a systems approach to sustainability. There are a number of ideas in the systems literature, especially in the emerging area of complex adaptive systems, that we believe are of fundamental importance in developing a true understanding of sustainability and of the related areas. This report is for those who want a general introduction to these more fundamental principles. We have sought to establish the general principles and concepts, then look from that particular perspective at some of the current issues and dilemmas in economics, philosophy, ethics, politics and so on, pointing out some of the implications for a number of areas. Some of the issues are revisited in several chapters, and discussed from various points of view. This is because we believe that a transition to a more sustainable way of life will require a significant change in the way in which problems are perceived, defined, and 'solved', and that this change must be away from a *closed systems* perspective, in which there are simple definitions, fixed concepts and ultimate solutions, to an *open systems* perspective, in which both problems and solutions are multi-dimensional, dynamic and evolving.

The book is written as an introductory text to the systems perspective for readers at approximately undergraduate level. It may also be helpful to those who want more substance than is provided in the more applied or popular literature. It is not written in standard academic format. There are no footnotes, and references have been kept to a minimum and keyed to the text by numbers, rather than given in full. This is to help the interested general reader follow the flow of ideas without too many distractions. The references given indicate further reading for those interested in particular topics.

1

Introduction

Such is the history of it. Man has been here 32,000 years. That it took a hundred million years to prepare the world for him is proof that that is what it was done for. I suppose it is. I dunno. If the Eiffel Tower were now representing the world's age, the skin of paint on the pinnacle-knob at its summit would represent man's share of that age; and anybody would perceive that that skin was what the tower was built for. I reckon they would, I dunno.

Mark Twain, *The Damned Human Race.*

The brief history of humanity

The Earth is about four and a half billion years old, somewhere between one quarter and one half of the age of the universe. It is the only planet in the universe currently known to support life. Life began on this planet nearly four billion years ago, so the story of life on Earth is only a little shorter than that of the planet itself.[1]

The human species is of relatively recent origin. Humans have existed for some 0.005 per cent of the time during which there has been biological activity on the planet, a ratio approximately equivalent to one day in a 55 year lifespan. The initial divergence from other apes occurred some 7 million years ago. The precursor *Australopithecus* lived from 7 to 2 million years ago. *Homo Erectus* evolved, via *Homo Habilis*, some 1.7 million years ago. Our immediate ancestors, *Homo Sapiens*, evolved from *Homo Erectus* possibly as little as 200,000 years ago. From about 100,000 years ago *Homo Sapiens* occupied parts of Africa, and the warmer parts of Europe and Asia.

It is not clear at what stage we became a cultural species. It has been suggested that there is evidence for cultural behaviour from as far back as 60,000 or even 100,000 years ago. However, the earliest unambiguous evidence for sophisticated cultural behaviour, including a technology of tools and weapons, burial of the dead, fertility worship, paintings, sculptures and so on dates from some 40,000 years ago, as modern man, *Homo Sapiens Sapiens*, spread across Europe and replaced Neanderthal Man (now generally considered to be a member of the species *Homo Sapiens*). Humans reached Australia some 35,000 years ago, North America perhaps 20,000 years ago, and had spread across most of the ice-free world by the end of the last ice age, some 15,000 to 12,000 years ago. The first domestication of plants and animals happened some 12,000 years ago, and there were farming communities in various parts of the world by some 8,000 years ago.[2] Some of these villages grew into the first small cities some 6,000 years ago. In comparison to the duration of life on Earth, therefore, contemporary human civilisation is of very recent origin.

The origin of civilisation

The original development of permanent human settlements was probably rooted in both functional and cultural needs. All species are faced with the functional problems of survival and reproduction, which has led to the evolution of some structurally similar solutions. There are situations in which some form of social organisation or other form of cooperation makes it more likely that the individuals concerned will survive and reproduce, and that their genotype will therefore survive. In such cases, the process of evolution will tend to select for cooperative modes of behaviour. Certainly, humans are not the only social species. It is probably inappropriate to try to draw any moral implications from this, or to conclude either that humans are essentially selfish or that humans are essentially social and cooperative. Such arguments are generally based on a simplistic genetic determinism. The process of evolution has no moral connotations.

However, the first semi-permanent settlement sites may have also had important cultural significance from the outset. It has been suggested, for example, that the core of a group's territory would have had great symbolic importance as the place where they buried their dead, had their most sacred sites, and carried out their most important rituals.[3] It is probable that the development of humans into a fully social and cultural species, with sophisticated political and economic modes of organisation, was primarily achieved through the medium of the proto-city, which made such social differentiation and organisation possible. Some of the complex patterns of relationships, functions, and purposes that constitute modern civilisation today probably existed, in an embryonic form, in the neolithic villages 11,000 years ago, and evolved to meet the more complex demands of larger settlements as villages grew into the first small cities some 6,000 years ago.

As humans have developed into a cultural species, we have acquired the ability to regulate the pattern of interaction between members of the community and between the community and its environment via socially transmitted information rather than biological feedback processes. The concepts of ethics, rights, responsibilities and codes of behaviour, for example, probably originated in the functional need to manage both the internal group dynamics and the interaction between the group and its wider natural and social environment by controlling individual and interpersonal behaviour. Such concepts would have been first expressed as customs or mores, group-based behavioural codes that would influence the behaviour of individuals in ways that would aid the survival of the group. Such group-based customs have been developed and extended, over thousands of years, into philosophical, moral, legal, governmental and other codes and systems of behaviour, as well as conceptions of organised morality in the form of religious proscriptions.

The increasing numbers in settlements, with the consequent concentration of demand and technical and religious power is also likely to have generated early political conflicts and structures. The land-use and other resource demand generated by neolithic settlements, for example, probably created conflicts of interest with residual palaeolithic migratory people. This conflict is reflected in various religious and mythical forms, such as the story of Cain and Abel in Genesis, and Lugubanda and Dumuzi in Mesopotamian legend.

As settlements grew, councils of village elders were replaced by more differentiated political structures, with the first emergence of castes of warriors, priests, and kings. Of course, the development of increasingly sophisticated social organisation imposes its own demands. There are new opportunities for the interests of the individual to clash with those of the group, and such conflicts must be managed in such a way as to limit the damage that might otherwise ensue. Thus the early cities probably gave rise to the first political and economic forms and structures.

The death of civilisations

There have been a number of occasions in human history in which societies have failed to meet their challenges, and have ceased to exist. In some cases this has been due to natural disasters, where events of non-human origin such as earthquakes, tidal waves, and volcanic eruptions have made life impossible. In some cases disintegration has been caused by invasion and the destruction of pre-existent social, cultural, and economic systems. There is also a third category of cases, in which societies have damaged their environmental support systems to the point at which their demands or their numbers exceeded the reduced carrying capacity of the area. The collapse of the civilisations of the Old Kingdom of Egypt around 1950 BC, the Sumerians in 1800 BC, the Maya at about 600 AD and of the Polynesians of Easter Island at about 1600 AD, for example, all appear to be of this type.[4] Although the circumstances were varied, these instances of social disintegration appear to have been at least partly caused by a pattern of demands that degraded the resource base and eventually exceeded the available flows of resources, with a critical conjunction of external environmental change with various social, cultural, political or economic circumstances that resulted in a failure to develop an adequate response to the changing situation and a consequent inability to achieve a more sustainable balance. Another common feature of these examples is that while the final stage of social disintegration was relatively brief, the causes of decline developed slowly and so may not have been sufficiently obvious until rather too late.

S. J.
GOULD

While there have been some societies that have exhibited a relatively high degree of stability over lengthy periods, a 'profile' of human society over the 40,000 years of our existence as a cultural species would show points of relative equilibrium, punctuated with periods of more rapid change. All systems have some ability to resist perturbation, or they rapidly cease to operate as systems. However, no living or social system is static. Cultural values change, political and economic systems develop, expand, and collapse, and environmental systems evolve, grow, and perish. It is therefore more accurate to say that societies have periods of relative stability and instability. A society in which numbers and demands are low relative to available resources, and in which the area ecology has absorbed the human impact while retaining the same essential characteristics is more likely to be ecologically stable than one that does not have these attributes, while a society that has a strong consensus on the distribution of power and resources is more likely to be politically stable than one that does not have such a consensus. Such attributes always vary over time, so that a society could become politically or ecologically unstable when some key element of political support fails or the availability of some critical environmental resource falls below the necessary minimum.

Those societies that have achieved periods of stability do not generally appear to have planned explicitly for sustainability. Customs or policies that had the effect of maintaining a particular human-environmental balance, for example, would have made that society more likely to survive. Those groups that did not have such customs would be less likely to survive. As many social structures in pre- or non-literate societies are largely unrecorded, most of the available evidence (in the form of surviving societies) would be on relatively successful models of behaviour. This may be partly responsible for giving rise to the idea of the 'noble savage', living in harmony with nature.

Throughout its brief history, the human species has extended its influence and control, expanded into a wide range of environments, and changed or displaced a number of other ecological balances. This human-environment interaction has always been a process of dynamic evolution. The recent development of formal science and technology, and the consequent extension of human influence, has accelerated the actual and potential rate of change. Our numbers are currently increasing rapidly, and our demands have increased and become more diverse.

There are now a number of issues that suggest that we should start planning explicitly for sustainability, for perhaps the first time. The alternative may be that current human society will undergo the kind of social and economic dislocation and collapse that a number of earlier societies have already experienced, and that the situation will in that way self-correct (in the sense that demand will then fall to within carrying capacity), but at considerable human cost. Perhaps the sole consolation would be that at least this time the causes of failure would be relatively well documented for some future historian.

The real difficulty is that human society now appears to be facing a global *problematique*, a complex of interacting complex problems. Many countries are undergoing major political and economic transformations, and balances of power are suddenly fluid and shifting. Political concepts are changing as well. Nationalism is resurgent, yet the concept of the nation state itself is being questioned in the emerging economic groupings. Previously dominant economies are manifesting serious structural weaknesses, other economies are expanding at rates that cannot be sustained. New technologies are developing, old ones are relocating to other parts of the world. Patterns of resource demand and pollution output are changing accordingly.

More fundamentally, it now appears that human activity can, potentially, affect the global ecology in ways and to an extent not previously thought possible. Global warming, ozone depletion, soil erosion, deforestation, desertification, and species extinctions are all indicators of the extent to which human activity is now altering the conditions for life on Earth. There is now concern that current and projected human demands might exceed the mineral and biological flow rates that the planet can yield without adverse consequences, such as ecological, social, or economic disruption. Similarly, there is concern that the flows of wastes generated in meeting these demands could exceed the capacity of the global ecology to absorb these wastes without similarly adverse consequences. Some of these problems might, if unchecked, eventually cause widespread damage to human society. Some could conceivably threaten the survival of the human species. There is currently considerable uncertainty as to just how realistic these scenarios are. This means that there are no definite grounds for either alarmism or complacency. However, given that the worst-case scenario is very severe, prudent action would probably be wise. Some of the more obvious environmental problems are reviewed in Chapter 4. It is important to note that the problem is not one of absolute shortages of energy, resources, or pollution absorption capacity. The problem is in the pattern of interaction and usage. It is, in effect, a problem of poor management.

There is probably no currently possible human action that could threaten the existence of life on this planet. There are few conceivable human actions that could achieve the scale of ecological disruption already experienced by this planet on a number of previous occasions, such as the mass extinction at the Cretaceous–Tertiary boundary 64 million years ago that eliminated some 50 per cent of all marine species, or the Permian–Triassic boundary event 250 million years ago that may have eliminated some 95 per cent of all species.[5, 6] Human action is, however, having a significant impact on the global ecology, sometimes unknowingly and to little advantage. It is probably within current human capability to pursue these actions to the point where severe damage to human society would result, possibly even to the point where it could shorten the potential lifespan of the human species.

Of course, the ultimate fate of all species is extinction. More than 99 per cent of all species that ever lived are extinct. Species have very varied lifespans, and while the blue-green algae have been here for about 3 billion years, the typical lifespan is very much shorter. The potential lifespan of the human species is unknown, but not infinite. However, premature extinction would be regrettable.

The changes in the global ecology indicate that we need to become more aware of the consequences of our actions, and to start to manage our affairs more consciously than has generally been the case in the past. This may mean that it will be necessary to evolve new political and economic structures and decision-making mechanisms in order to respond to these emerging global environmental demands. However, as indicated earlier, we may have to do so from a position of relative political and economic instability. This is likely to be a challenging process.

Defining problems and finding solutions

If these issues can be adequately addressed by making relatively minor changes in the ways in which society meets its social and economic goals (as currently defined) a comprehensive re-evaluation of these goals would be redundant. If, however, an adequate response would entail more fundamental social and economic restructuring, it would become more likely that the goals themselves would have to be re-examined. Later chapters in this book will argue that the number, complexity, and inter-relatedness of the issues now indicate that a strategy that consists of relatively unconnected adjustments to social and economic policies and means is less likely to be successful than a systematic attempt to construct socio-economic systems that engage and interact appropriately with the ecological systems of the planet. Such an approach would include the evaluation and reappraisal of many social and economic goals.

It is also clear, for reasons that are reviewed in later chapters, that many existing organisational, political, and economic concepts and structures are probably now inappropriate and unhelpful. It is unlikely that the necessary structures for international coordination, for example, will be evolved without some degree of organisational and political transformation. This in turn is unlikely to happen without a parallel evolution of the cultural and psychological concepts on which political and economic structures are ultimately based.

This is why any analysis, to be adequate, must include the relevant environmental, political, economic and socio-cultural factors. The sustainability of the human species can only be defined, ultimately, at the level of the interaction between the entire complex of human systems and all directly implicated environmental systems. To understand sustainability therefore requires some understanding of the behaviour of systems in general and of human and environmental systems in particular.

There are a number of definitions of sustainability currently in use. There is some consensus that a transition to a sustainable way of life means taking steps to try to reduce the risk that environmental and related problems will seriously affect or jeopardise the human species at some future time, and thereby to ensure that future generations have a reasonable prospect of a worthwhile existence. The question of sustainability is, therefore, one of enlightened self-interest. It requires finding ways in which the human species can live on this planet indefinitely, without compromising its future.

Sustainable does not mean static. Every ecology, including the global ecology, is a dynamic interaction of interlocking cycles. A stable ecology, which is at a point of balance between interacting forces, continuously processes and cycles energy, nutrients, and other resources. The point of balance itself changes over time. The history of life on Earth has been one of contingent evolution. This has involved both slow incremental change and phases of rapid transition. Change and evolution is inherent in this process.

All species interact, change, and co-evolve with their environment. The human species is no exception. We are exceptional, however, in our ability to modify consciously some elements of the pattern of our interaction with the environment. It is no longer possible, given the current extent of human activity, to avoid making these management decisions as to how we wish to interact with the planet. For example, a decision not to cull the Scottish red deer, given that important natural predators no longer exist, is now a management decision, just as much as a decision to cull the deer. Every permutation of all decisions of this type has ecological consequences. Similarly, a decision to disregard information on current global ecological trends is a management decision, as much as is the decision to attempt to achieve some particular human-ecological balance.

While many would now agree that there is a need to find ways of living that are sustainable, the discussion to date has yet to generate a fundamental explanation that spans the issues and provides a coherent reason and direction for social change. The authors believe that the best prospect for an effective response to the global crisis lies in developing an analysis that can offer both; one that can both justify and inform a long-term, integrated and coherent strategy for change.

Such an approach is slightly unfamiliar in the UK. This is because Britain has a general philosophical ethos that is more empirical and instrumentalist than the mainland European intellectual tradition. Pragmatism and practicality are valued, abstract analysis is distrusted. This ethos underlies a situation in which specific and concrete responses to problems are valued (as indicating pragmatism and practicality) while more abstract appraisals (which might lead to more fundamental and comprehensive solutions) are often seen as being less useful.

This approach has both strengths and weaknesses. Its strengths are that it emphasises specific and applied action and, in the short-term, enables progress at minimal cost, minimises the creation of unnecessarily large bureaucracies, and incurs a lower risk of misapplied time and effort. Its main weakness is that it is non-abstract, which means that underlying similarities in particular problems are often overlooked. This in turn means that it encourages ad hoc responses to immediate pressures, which may seem pragmatic while masking a failure to recognise, let alone address, the underlying causes. It thus encourages the creation of multiple solutions to what may be aspects of a common problem, and these solutions may, collectively, be less efficient and more costly than a single comprehensive solution. It is even possible for such multiple solutions to be in mutual conflict, thereby rendering the whole approach even less efficient.

This can be seen in the UK's approach to environmental protection to date, which has generated a large number of regulatory agencies with areas of responsibility allocated on inconsistent bases. Some agencies are allocated responsibilities in terms of the *substance* involved (such as the National Radiological Protection Board, which has responsibility for radioactive materials), some in terms of the *medium of disposal* (such as the River Purification Boards), some in terms of the *geographical region of responsibility* (such as the local government responsibility for regulating the disposal of certain kinds of wastes).

This situation has arisen, of course, for historical reasons. Britain, as the first country in the world to industrialise, was the first to encounter some of the problems associated with industrial pollution. Responses to the problems were, in general, developed on a case-by-case basis. There was no body of experience elsewhere, at the time, on which the UK could have drawn, and which might have allowed the UK to anticipate some of the problems associated with a necessarily fragmented approach. The weakness of such an approach is seen in the fact that it has, in this case, fostered the creation of a number of agencies, some with overlapping areas of responsibility. The approach that these agencies take, in terms of the frequency of inspection, severity of penalties and so on, is not uniform. This has created scope for some firms to switch between waste disposal media (from, for example, river disposal to land fill or combustion) in order to shift to a less stringently regulated or well-policed medium, which can allow such firms to 'fall through the net' and evade effective regulation, and result in problems being displaced rather than solved.

It seems unlikely that we will succeed in the far more complex task of achieving sustainable development without a coherent basis for programmes of action in all relevant areas of life. The vital task, if we are to devise such a common basis for action, is to find a way to integrate the critical dimensions of the debate, to bring in the essential scientific, socio-economic and philosophical information and perspectives, and to develop a rational and comprehensive strategy that can generate practical and effective policies.

The role of natural science, social science and philosophy

Different disciplines and schools of thought have different perspectives to contribute to such an understanding. Some of the more obviously significant contributions come from the natural and social sciences, and from philosophy.

Natural science

Scientific research is essential if we are to develop our understanding of the behaviour of the geological, ecological, biological and other processes that shape the global environment, to monitor change, identify trends, and predict possible outcomes. It is important that such research is done on a properly sceptical and impartial basis, and is kept as free from political and other constraints as possible.

Some have argued that science must become less sceptical and more committed to the cause of environmental protection, but this argument is usually based on a confusion between the role of scientists and the nature of science. Scientists, being human, are subject to the same unconscious prejudices and biases, failings and moral conflicts as everyone else. The essence of science itself, however, lies in the attempt to discover what is true, regardless of other considerations. Critical methodology is our only safeguard against the human ability to deceive ourselves and others.

There are, however, genuinely important questions, especially in this context, as to how scientific knowledge is understood and used. Scientific knowledge is partial and subject to refutation (although the degree of confidence in results does vary markedly between different domains), so that any policy decision based on scientific research has to be made on the basis of an estimate of risk. Additional decisions must be made, therefore, as to how to manage and apportion that risk. A decision to accept a slightly higher risk of adverse consequences resulting from some long-term environmental problem could have the effect of redistributing the costs and benefits of remedial action in favour of the current generation and to the disadvantage of some future generation. This raises a number of political, economic and ethical issues. People are concerned, not just about the degree of risk, but also about how the risk is distributed.

Such questions are endemic in the sustainability debate. It will be argued in several later sections that it is unlikely that there will ever be only one route to a more sustainable way of life, and that there will always be political and economic choices of this kind to be made in meeting any given scientifically-determined target. Thus the debate is in part about science, and in part about politics and economics.

Social science

Social, economic and psychological research is essential if we are to determine which economic or other policy tools would achieve desired ends (that society should remain within certain total resource consumption or pollution output limits, for example) with the maximum economy of means and with the minimum adverse effect on other social and economic goals. Socio-economic research is also necessary to estimate the likely redistributive and other social consequences of any such policy decision.

Further research will be required to develop techniques for assigning and incorporating environmental values, where appropriate, into economic and related decision making. It is now clear, for example, that many direct and indirect environmental inputs into economic processes have been undervalued. Forests have been felled for timber with no allowance made for their function in maintaining biological diversity. The pollution absorption capacity of the atmosphere has been partly 'consumed' by the generations who have enjoyed the benefits of cheap fossil fuel, and who have not met the full cost of this consumption. In general, inputs that are undervalued will tend to be used to excess. One possible response, therefore, is to use taxes or charges to increase the cost of these inputs.

A number of economists have developed techniques for assigning and incorporating environmental values into increasingly sophisticated cost-benefit analyses. These approaches, thus far, have had a number of difficulties. Technical problems with the quantification of variables and the calculation of values (especially given existing market distortions), and the generally poor reliability and validity of valuation methods have meant that results have varied very widely between different studies. There are, in addition, more fundamental doubts about the assumptions that underpin the orthodox neoclassical approach, including the exogenous formation of human preferences, the concept of value, and the concept of utility along with the theory of hedonistic associational psychology on which the concept of utility is based. Another problem for any attempt to find 'real' market values for environmental inputs is that there is no evidence that human preferences expressed, for example, through the market place relate to the ecological value or criticality (as opposed to the simple economic scarcity) of forms of environmental inputs.

Taxes, charges, and subsidies can, of course, be useful tools for changing aggregate behaviour, irrespective of whether they relate to any 'real' market values or any external value at all. Another approach, therefore, would be simply to set them at the level empirically found to be necessary to bring about the desired change in behaviour.

The fundamental dilemma, of course, is whether society should accept external environmental limits, and modify its behaviour accordingly (which means accepting some limits on human welfare, as conventionally defined), or whether to assert the primacy of human values and human welfare (so that any other factors can only be said to have value insofar as they can be translated into human terms). Such questions, of course, mean that there is a philosophical dimension to the debate.

Philosophy

Philosophical and psychological analysis is essential in order to elicit and reveal the mix of rational and irrational assumptions that underlie human decision-making processes, and to explore ways in which non-local, long-term and diffuse relationships between actions and consequences can be brought into a practical ethical framework for decision making.

Many of the issues in the sustainability debate concern us directly. If it were decided that the resources of the world should be shared out on a more equitable basis, for example, individual standards of living around the world would change dramatically. Such factors mean that our personal political and moral perspectives tend to influence our positions in the debate. If we, at some level, are trying to make genuinely objective decisions, it is far better that such perspectives are stated, rather than concealed within a methodology. Good decision-making procedure requires that implicit value judgements be made explicit, so that more of the actual decision-making process becomes visible.

This is particularly important because many of the decisions that will have to be taken in a transition to a more sustainable way of life will not be between clear-cut 'good' and 'bad' alternatives, but will involve choosing between options that have different benefits and different costs, for different people at different times. Many of these decisions will have to be made on the basis of assumptions and estimates. This will mean making some fundamental choices, such as whether to err routinely on the side of caution, by imposing stringent burdens of proof of safety on developers, or, at the other extreme, permitting any development not actually proven to cause serious harm. Such decisions often entail making value judgements as to what the priorities of society should be, so these questions then become part of the debate.

Similarly, a decision made in the UK today might affect people that live in distant parts of the world, or who are not born yet, and the link between the decision and the effect could be indirect and diffuse. It would be difficult, for example, to trace the environmental and other costs of an individual decision to buy a new car in the UK, when these might include a contribution to carbon loading of the atmosphere, global warming, rising sea levels, and loss of cropland, all of which are likely to have different implications for different parts of the world and for different generations. This means that there are moral judgements to be made as to the extent to which we should assume a general responsibility for decisions and actions that could affect people in these ways.

In practice, therefore, environmental questions are inextricably interlinked with social, economic and cultural values. Economic systems determine the rate and route of flows of energy and resources from the environment into patterns of human use, and the rate and route of flows of waste energy and materials from human economic operations back into the environment. These economic systems are, in turn, imbued by cultural values, and underpinned by social and psychological models that influence the way in which people understand their options and make their choices.

The scientific, economic and philosophical approaches to the *problematique* clearly have a vital contribution to make to our understanding of the issues. Yet the fact that they all do in itself indicates that they are all concerned with one aspect of the *problematique*, and that if we wish to understand the *problematique* itself, we need to have a way of integrating these various contributions.

Integrating information from different domains

The need, of course, is to have some way of incorporating information from such different domains into a single decision-making process. One approach is to try to map all the relevant information into one domain. This underlies the attempt in environmental economics, for example, to assign values to ecological and social phenomena, so that they can be brought into one cost-benefit analysis and one decision-making framework. There are three important problems with this approach.

❑ It resolves a great deal of uncertainty (and possibly even true indeterminacy) in these other domains in this process of translating and mapping the information. It provides the maximum simplicity in the final decision but, by exactly the same token, entails the maximum loss of information. Information loss itself, of course, is not invariably a problem. The real concern here is that critically important information may be lost.

❑ Such an approach will not, in general, contribute to any real understanding of the dynamic interaction of complex ecological and economic systems. It may even serve to distract from the approach needed to develop a more complete understanding.

❑ The third problem concerns the way in which information is used in the real world, and the political issue of the relationship between information and power. Any decision-making procedure that uses a single index, such as cash or welfare, must assign values and thereby weightings to a range of diverse factors. The choice of methodology for so doing is usually the most significant part of the entire decision-making process. This, in practice, is usually invisible by the time the data has been processed and the resultant numbers come forward for the final part of the decision.

Systems theory

A systems approach, in this context, has much to offer. It provides a multi-dimensional framework in which information from different disciplines and domains can be integrated without being forced into a one-dimensional mapping. A systems approach to sustainability entails considering the various agents interacting in the world as systems. This involves invoking general principles concerning systems to make inferences about likely and actual interactions between the systems under consideration. These principles can also be used to analyse the observed patterns of interactions between systems.

The world itself can be thought of as a very large and complex system, which contains complex subsystems such as ecological and biological systems, weather systems, and human social and economic systems. Subsystems such as the human economy are open systems (systems that exchange energy or resources with their environment). The economy is a highly open system, as it interacts with many other systems in a great many different ways. This becomes important when developing policy to regulate a pattern of interactions between such open systems.

In general, complex systems generate outcomes that depend on numerous interactions. As a result, many complex systems are highly sensitive to the precise starting conditions and loading of factors. Some kinds of complex systems can, in addition, behave chaotically in certain circumstances, which means that a small change in some control variable can result in large and disproportionate changes in the behaviour of the system. For these reasons, models of many types of complex systems are inherently probabilistic and limited. Global subsystems, such as socio-economic and environmental systems, are themselves complex systems, and the pattern of interaction between them is yet more complex. It is unlikely that any simple models will be able to capture this range of behaviour. For example, simple models often fail to predict effects beyond system boundaries and thresholds, at which the behaviour of the system itself may change qualitatively.

As the earth system evolves over time, the environmental factors that shape the outcome of the evolutionary process also change. This alters relative selective pressures and the balance of evolutionary and adaptive advantage. The current system state provides the essential conditions in which the human species evolved and can survive. All system states change over various timescales. There are critical points in system state changes beyond which any given subsystem (such as a species, for example) will not be able to go, and when these thresholds are passed, the subsystem concerned ceases to exist. This means that irreversible losses will naturally occur in the course of time. For any one species, such as the human species, some of the dimensions of the system will be more important than others. Change on some dimensions would not be significant, while change on others would be fatal. This means that current and future irreversibilities will affect the human species to different degrees. The continuing survival of the human species depends on our avoiding an irreversible transition on a critical dimension.

The size and complexity of the earth system indicates that there could be, at any one time, a very large number of potential development paths and possible outcomes, a smaller subset of which would be relatively sustainable for the human species. This in turn indicates that there could be a number of states that are sustainable in varying degrees, that there may be a number of ways to reach such states, and that there will therefore be more than one possible policy for a transition to a more sustainable way of life. This may allow the simultaneous achievement or integration of other philosophical, political, or economic values, but it is important to be aware of these distinctions.

We cannot assume, however, that we currently understand the earth system well enough to be able to identify all of our critical dimensions. One essential component of the task, therefore, is to identify the important dimensions, and to develop the means to monitor movement on these dimensions.

This in itself would not be enough, however, and a further key component of the task is to develop a model to translate that information into a form that can be used in decision-making processes.

Sustainability Assessment Maps

The need, therefore, is for a decision-making process that can accommodate change in a number of non-equivalent dimensions. The decision-making model selected for more detailed discussion in this report is based on the use of a technique called positional analysis and a related graphical tool called a sustainability assessment map (SAM). This is a relatively simple and intuitively fairly obvious way of representing change on a number of dimensions simultaneously, which helps to identify some of the inevitable trade-offs involved in any significant change or development. The technique represents a reasonable compromise between the need for more open, transparent and accessible decision-making and the need for better informed and more focused discussions.

A SAM consists of a diagram in which each of the critical dimensions in a compound problem is represented by an axis. Measurements of change or indications of priorities are then mapped onto these axes. The resultant profiles can be differentiated, which highlights the trade-offs inherent in each possible choice. This purpose of this approach is to emphasise rather than conceal trade-offs, and to do so in a way that is as accessible and intuitively obvious as possible.

When making a strategic development decision, for example, the first step is to identify the critical axes of change. These could include costs, profits, numbers of jobs, environmental impact, types and quantities of natural capital required and so on. Each of the main development options is then assessed on the same basis and scored on all of the axes concerned. The scores are then differentiated, and the results displayed in the SAM graphic. Each option can then be compared in terms of its overall profile. This compare-and-contrast procedure clarifies the differences between two or more development options in terms of the complete profile of costs and benefits obtained in each case and the overall balance of advantages and disadvantages that each option has with respect to each of the others. A SAM can also be used to show movement on these multiple indices in order to demonstrate change over time.

Contents of this book

This book, then, reviews the concept of sustainability, explains what the systems approach is, what it means for our understanding of the problem of sustainability and the methods by which we address it, reviews some of the relevant social, economic, and environmental factors, and introduces some of the ideas and tools needed to translate the concept into economic, political, and social decisions.

It also suggests a set of criteria to allow political and economic decision-makers and planners to assess their strategies in terms of their impact on sustainability, and puts forward a series of proposals for specific actions to enable a general transition to a more sustainable way of life.

2

General Systems Theory

The nature of science

Science is a way of acquiring testable knowledge about the world. The knowledge which science provides is always provisional and probabilistic. All theories are approximations to the truth, with a certain domain of validity. Science has a number of defining characteristics. Of these, three are particularly important. These are as follows:

❑ *Replicability.* Scientific knowledge must be as objective as possible. This means that different observers must be able to replicate results and thereby verify the observations.
❑ *Refutability.* It is impossible to carry out all possible experiments, as there is not enough time in the lifetime of this universe. It is therefore important to decide which experiments to do. Good experiments are those that help to decide between competing hypotheses. Although many scientists, being human, tend to prefer to have their theories corroborated, good scientific technique consists of trying to refute explanations. This is because, while a conclusive refutation is decisive, corroboration is not. The fact that the sun has risen every morning so far does not prove that it will rise tomorrow, whereas one convincing failure to rise does prove that the sun does not invariably rise in the morning.
❑ *Reductionism.* For a number of reasons, science is generally reductionist. First, and most important, the real world is vast and complex, while humans are finite beings. This means that complexity must be mapped onto a relatively simple model that humans can encompass and comprehend. A second reason, which is directly related to the first, is that people generally find minimal and economic explanations more coherent and convincing. A third reason is that reductionist analysis has been, so far, the most successful explanatory technique available.[7]

The problem of complexity

Complexity poses some of the most difficult problems for contemporary science. In studies of large-scale phenomena, such as populations, there is usually sufficient diversity for the data to be treated statistically. The situation can be reduced to simplicity by dealing with aggregates and averages, and by treating units as equivalent and interchangeable. Small or well-defined problems, on the other hand, can be treated analytically. Extraneous and insignificant factors can be excluded, and the situation reduced to its essentials. Thus studies of subjects that are large enough to be effectively random or small enough to be precisely defined can be studied with a standard reductionist approach.

Between these two extremes, however, lies a third group of phenomena that are too diverse for analysis and too structured to be random. In such cases the standard reductionist approach is less useful.[8]

The standard approach essentially consists of disaggregating the effects of each of the elements in a situation, and thereby identifying the discrete contribution of each element (this is true whether the element concerned is in reality a discrete unit or an abstraction such as an aggregation of similar units). The standard approach depends on the assumption that the separation does not affect the operation of the parts, so that the cause and effect relationships contained in the subject under study can continue to be adequately represented.

There are cases, however, where it is very difficult to ensure that the cause and effect relationships are being adequately represented. This kind of problem arises when the elements in a situation are being organised in some way, so that they are behaving in that specific context in a way that cannot be predicted solely from knowledge of their individual behaviour. Separation, in such instances, does affect the operation of the parts.

Such organisation can be seen in living systems, for example, where it is reflected in the hierarchy of organisational structures and control processes from atoms to ecologies, via the intermediary levels of chemicals, molecules, cells, organs, organisms, social groupings and species. Behaviour at each lower level cannot be easily explained without reference to organising functions at higher levels. The precise arrangement of the organic bases in DNA, for example, is crucial in biological terms, although most possible arrangements would be compatible with the laws of physics and chemistry.

Each level in a hierarchy of organisational structures and control processes is significant in a number of ways. One is that there are important functional differences and behavioral discontinuities between levels. As a general principle, higher levels exert controls on lower levels by promoting or constraining various possible outcomes. Another is that there are various physical differences between levels. In many systems, for example, the energies between the units vary greatly between the different levels. The binding energy between nuclear particles is about 140 million electron-volts (eV), between the atoms in a molecule about 5 eV, while between molecules about 0.5 eV.

We recognise this hierarchical ordering of the world into levels of complexity in various practical ways. The underlying logic of the traditional division of the academic disciplines, for example, from physics to economics, via chemistry, biology and psychology, is that each concerns itself with a tier in this hierarchy of nature. Each discipline has therefore developed an appropriate methodology and epistemology. It is important to remember, however, that while the divisions between tiers are highly significant in some regards, they are less so in others. There is continuity too, in that the same elements are physically present throughout the hierarchies. A given element can, in fact, operate both in a control function and in a dependent function in different circumstances.

It is important, therefore, to ensure that we never allow our thinking to become unduly constrained by disciplinary boundaries to the point where we might fail to observe deep underlying continuities in terms of organisational processes and structures. Progress in science and in our understanding of the world is often made when attempts are made to bridge into the domain of other disciplines.

Scientific methodology and definitions of proof, however, cannot be uniformly applied throughout the spectrum from physics to economic behaviour. Situations appear to become more variable, less controllable and less predictable as complexity increases. System behaviour appears to become increasingly unrestricted. The response of an individual human operator in an economic system, for example, is less predictable than the response of a given molecule in a chemical reaction.

This means that knowledge itself becomes less certain. While we can be relatively certain about the laws of physics, it is probable that there will never be similarly comprehensive laws in the social sciences. One practical problem, for example, is that events in the social world do not replicate exactly. There are, clearly, patterns of events that replicate, but this calls for some degree of judgement as to what degree of overlap constitutes a genuine replication. This means that the strict methods of the natural sciences cannot be similarly applied in the social sciences.

As discussed in Chapter 13, it is important to note that this is not an argument between reductionism and holism. The world is vast and complex, and the human ability to process information is limited. All models of the world are reductionist, therefore, as information loss must be accepted in order to gain simplicity and clarity. The need is for an intelligent and sophisticated reductionism.

General systems theory was developed, in response to this perceived need, to provide a unifying analytical and explanatory framework throughout the hierarchy of nature. It provides a tool for integrating the contributions of different disciplines. A systems approach, in essence, entails considering the various agents interacting in the world as systems. This approach involves invoking general principles concerning systems to make inferences about likely and actual interactions between the systems under consideration. These principles can also be used to analyse the observed patterns of interactions between systems.

The nature of systems

A systems approach involves placing as much emphasis on identifying and describing the connections between objects and events as on identifying and describing the objects and events themselves.

A system, formally, is a set of components that interact with each other. Changes in one component will induce changes in another component, which may in turn induce change in a third component. Any one interaction of this kind is causal and directional. Many such interactions can be linked together in chains of cause and effect relationships. A given component can often, in practice, operate both in a control function (causing change in another) and in a dependent function (being changed by another). This is called *multi-factoriality.* Chains of cause and effect relationships can intersect themselves. This means that a component can start a sequence of causes and effects that eventually loops back, so that each of the components in the loop indirectly influences itself. This is called a *feedback* loop. A system can contain more than one feedback loop. It is also possible, in a system that contains multiple feedback loops, for some or all of these loops to intersect each other, with the components constituting the loops acting in turn in control and dependent functions. The behaviour of a given component, in such a set of relationships, is the outcome of multiple competing factors. A feedback loop is called *positive* when the input back to the initiating component is excitatory, that is, makes that component more likely to initiate a further similar event sequence, and *negative* when the input back to the initiating component is inhibitory, that is, makes that component less likely to initiate a further similar event sequence.

Some of the more important defining characteristics of systems are emergence, hierarchical control and communication.

❑ *Emergence*. This means that at any given level of complexity, there are emergent properties that cannot be readily explained solely by reference to lower levels.
❑ *Hierarchical control*. Hierarchies are levels of relative complexity within a system, and hierarchical control refers to the imposition of new functional relationships by each level on the detailed dynamics of the level below. Controls can be positive (where certain actions are promoted) or negative (where certain actions are constrained). One of the challenges facing biological systems, in particular, is to optimise between excessive control (which gives little flexibility, a small behavioural repertoire and so a limited ability to respond to new circumstances) and insufficient control (which reduces the ability of the system to determine outcomes, and so incurs a higher risk that the internal processes themselves might drift beyond system limits and cause the system to disintegrate).
❑ *Communication*. This refers to the transmission of information in some form to effect regulation and feedback. Information must flow from the regulator to the regulated in order for the regulator to exercise

control. Information must also flow back from the regulated to the regulator if the regulator is to be able to monitor the compliance of the regulated and incorporate that information into its future programme. Positive and negative feedback loops are therefore the core of the process of communication. If the regulated has failed to produce an adequate response to the last signal, the regulator must send a similar or increased signal. If the regulated has produced an excessive response, the regulator may have to send a converse signal. In a central heating system, for example, where the thermostat is the regulator and the boiler is the regulated, an inadequate response from the boiler will cause the thermostat to continue to signal for an increase in output, while an excessive response from the boiler will cause the thermostat to signal for a reduction in output. The importance of such positive and negative feedback was first identified in a particular application of systems theory called cybernetics, which is primarily concerned with these processes of control and communication. The more recent extension of systems theory to include social and economic systems has required an evolution of the concept of communication, as it is the communication of meaning rather than information that is important in the world of human systems (see Chapter 13).

Open and closed systems

There is an important distinction between open and closed systems, which was first described by von Bertalanffy, the main founder of the systems approach.[9] This distinction depends on thermodynamics, so it is necessary to briefly review one of the most fundamental of all physical laws, the second law of thermodynamics. This states, roughly, that without the input of energy all systems tend to move from organised to disorganised states. This is a very profound principle, but one which is easily understood. It is obvious that of all the possible ways of arranging the components which make up any system, whether it be a flower or a computer, the configurations which are highly ordered and produce functioning systems are highly atypical, and that most configurations will be no more than a jumble of parts. The second law of thermodynamics says that, over time, even systems which begin as highly ordered will, with overwhelming probability, degrade into less highly ordered systems. While it is theoretically possible that all the scattered parts of a computer could reconstitute into a computer through random processes, in practice it takes energy to achieve this organisation. The amount of disorder in a system can be measured, and is termed the *entropy* of the system. Thus the second law says that the entropy of any system which is not receiving an energy input will, with overwhelming probability, increase over time. This is why things disintegrate, decay and die.

Life is an entropy-decreasing process. Living systems build, reproduce and create order. The reason that life can exist is that the Earth continually receives energy from the Sun. This solar energy is the essential motor which allows entropy to be decreased rather than increased, for flowers to grow, for people to build computers, and for order to be created out of disorder.

Now consider the distinction between closed and open systems. Closed systems have unchanging components. They will eventually arrive at an equilibrium, and tend to move towards a state of higher entropy. This means that closed systems usually reach a point at which no further change is possible for that system.

Open systems exchange flows with their environment. These flows can consist of materials, energy, or information. Open systems can reach a steady state, which depends on their being able to maintain continuous exchanges with their environment. This is what allows some open systems to create and maintain a state of low entropy. This means that some open systems can maintain their integrity as systems, although this must always be at the expense of an increase in entropy elsewhere.

All living systems are open systems. Of course, the environments in which living systems live are themselves never completely stable, so living systems must try to obtain reasonably stable flows from sources that can be changing over time. This means that living systems and ecological hierarchies of living systems must have processes of communication and control so that they can monitor and respond to, and in that way resist, the perturbations of a real-life environment. It is important to note that this does not have to be in any sense a conscious process. Effective control, in a changing environment, requires that systems have control mechanisms with a variety of response that can match the variety of the environmental information. This is sometimes referred to as the *law of requisite variety*.

It is, in fact, unlikely that complex life forms would exist today if they were not organised in this way. This is because the time required for the evolution of a complex form from single elements depends on the number and distribution of intermediate forms that are themselves stable, and because the age of the planet is such that only hierarchically-organised forms could have had time to evolve.[10] For example, imagine how long it would take to write the complete specifications for an aeroplane if it were to be assembled from the individual molecules of metal, as opposed to the subcomponents of engines, fuselage, wings and so on.[11]

Defining systems

There are a number of steps to go through when building a model of a system.

❑ The identification of the coherent elements of the system, and the definition of the principles of coherence. This can be quite difficult. Some functional systems are organised on an ad hoc basis, and can have different members according to what is required. Muscles, for example, are organised into different functional groupings appropriate to the task.[12, 13]

❑ The identification of the control mechanisms by which the system maintains its coherence, and the value ranges within which these operate. This can be complicated by the fact that biological and ecological systems are often characterised by redundancy, which means that they tend to use multiple pathways of control. For example, humans have three systems to maintain balance (visual, vestibular, and proprioceptive systems), all of which can function independently.

❑ The delineation of the system boundary. The system boundary defines the inputs and outputs to the system.

❑ The identification of any subsystems of the system, or supersystems (systems of which the system in question forms a subsystem).

There is no universally accepted classification scheme for systems, although it is generally agreed that all descriptions of systems should be as economical as possible.[14] Some general categories are living and non-living systems, concrete and abstract systems, open and closed systems. Other categories include basic, operational, purposive, and controlling systems. Others have grouped systems into five classes, as follows:[15]

❑ natural systems (atoms, planets);
❑ designed physical systems (machines);
❑ designed abstract systems (mathematics);
❑ human activity systems (political structures);
❑ transcendental systems (systems beyond knowledge).

The systems in these classes are of quite different types. In particular, designed and activity systems are different from natural systems. This is because humans are self-aware, and monitor their own behaviour, and because of the 'self-fulfilling prophecy' problem, where the knowledge of the prediction itself can become an input to the system, and alter the likelihood of the possible outcomes.

Modelling systems

In general, a good model will be economical, but have high predictive validity. This means that it should include all significant relevant factors, cover their range of values, and properly reflect their actual behaviour. It is also important that a model should exclude all insignificant and irrelevant factors.

One standard problem when modelling systems is how to decide what factors are relevant. It is often the case that values are not available for all of the factors that one might want to model, which means that one often has to work with those factors for which values are available.

One then has to find out how each factor affects the system. This too can be quite difficult. Effects can be non-linear, change their nature at thresholds, and be subject to delays (see p 25). Furthermore, in many real-life situations data is messy, which means that all of the above effects can be masked or diluted by the general variance.

There are four basic aspects of model validity:

❑ *Structural*. It is important to be reasonably sure that the model has captured the essential system structure, and that the structure of the model reflects the elements, interconnections and feedback loops present in reality.
❑ *Behavioural*. One important check is whether the model generally behaves in the same way as the real system, and manifests the same oscillations, thresholds, instabilities, changes, equilibria and so on.
❑ *Empirical*. The next step is to ensure that the model actually behaves like the real system given the same parameters and conditions.
❑ *Application*. The whole point of developing a model is that it should help to answer questions and thereby inform policy. A good model will therefore generate information that is in a useful form.

Models of linear systems usually attempt to provide a relatively precise specification of the system in a set of equations and so capture its behaviour. If the equations accurately reflect the structure of the system, it should be possible to model its behaviour with some precision. Most conventional economic models, for example, are of this type, typically involving hundreds of equations.[16]

One of the interesting features of models of non-linear systems, by contrast, is that the essential system dynamics can sometimes be captured in a relatively small number of equations – sometimes as few as two or three.

It is clearly important and useful to simplify and capture the essential behaviour of a system in this way. It allows us to develop our understanding, test our assumptions and estimate the effects of particular decisions and actions on the model before risking them in the real world.

Dynamic system behaviour

Very complex dynamic or adaptive systems, such as the weather, evolutionary processes or market operations, pose new kinds of modelling problems. It is very difficult to model and predict the behaviour of such complex systems.[17] We do not yet have all the necessary techniques and tools. In general terms, however, it is clear that the pattern of connectedness between the elements of a dynamic system is centrally involved in determining the behaviour of such systems.

Dynamical system behaviour is generally grouped into four classes, as follows:

❑ *Class 1*. Fixed, where the system is 'frozen'.
❑ *Class 2*. Periodic, where the system runs through a fixed cycle.
❑ *Class 3*. Chaotic, where the system is for all practical purposes nondeterministic. This can arise even in systems that are known to be

following a deterministic set of rules, when it is referred to as deterministic chaos. Whether the chaos is deterministic or truly nondeterministic, however, the detailed behaviour of such systems is effectively unpredictable.

❑ *Class 4.* The edge of chaos, where the system is in dynamic tension between flexibility and stability. This is a narrow but very important zone that is intermediate between class 2 and class 3 behaviour. Living organisms, for example, exhibit class 4 behaviour.

A system with few connections between the system elements is more likely to exhibit class 1 behaviour. If such a system is perturbed in one area, the effect is likely to remain local because the impact will not be transmitted far through the system. A system with a high degree of connectedness between the system elements is more likely to exhibit class 3 behaviour. Any change is more likely to propagate throughout the system, possibly to the point that the system becomes chaotic. Class 4 systems respond in a less predictable way. If a system at the edge of chaos is perturbed there may be very little response, or there may be a very large response, apparently out of all proportion to the size of the disturbance, depending on the precise internal condition of the system at the time.

Systems, by definition, have behavioural or other emergent properties that the components of the system do not, and which are not readily explicable with reference to the subcomponents. One very important feature of dynamic systems is that they can be ordered and stable. This stability can be an emergent property, a function of the interaction of individual elements in the system. An ecology, for example, can be maintained in a stable state by a dynamic interaction between the species that constitute that ecological system. Similarly, complex and adaptable but stable behavioural patterns are generated by the pattern of interaction of billions of neurons in the human brain. In a society, the aggregate behaviour of companies, consumers, and markets can be stable, even though the individual buying and selling decisions of the individuals that constitute the community cannot be predicted.

Complex adaptive systems

There is a very important and special class of complex systems that are termed adaptive. The unique feature that distinguishes these from other kinds of complex systems is that adaptive systems in some way interact with their environment and change in response to environmental change. Living systems, for example, are adaptive systems in some degree. They have a certain behavioural repertoire, even if this just consists of simple tropisms, which adapts to environmental changes. This adaptation could take place over a number of generations, or within the experience of one individual organism. Of course, some environmental change may be too rapid or too extensive for the organism or the species to adapt. This usually results in death or extinction. A given behavioural repertoire or schema can serve the purposes of the organism well under the circumstances in which it was evolved, but be maladaptive if the environment changes to too great an extent.

Systems terminology

This discussion of general systems theory, the behaviour of dynamic and adaptive systems, and the application of this perspective to natural and social systems is continued throughout this book. Chapter 3, in particular, goes on to consider complex adaptive systems in more detail. The following is an introduction to some of the basic systems terms and concepts which are used throughout.

This glossary has been developed from a number of sources, with especial reference to work by Checkland, Meadows et al and by Gleik.[15, 18, 19]

❑ *Attractor.* An attractor is a state towards which a system tends from other states. Once in an attractor, a system will tend to remain there until there is an external event of sufficient force or sufficient internal drift to dislodge the system from the attractor. Attractors can be strong (it takes a great deal to dislodge the system from the attractor) or weak (the system is easily dislodged). If states are imagined as points on a surface, attractors can be represented schematically with deep (strong) and shallow (weak) hollows in the surface into which a system can 'fall' and get 'stuck'. A system can move through a state space with multiple attractors, falling into one, remaining for a time, being dislodged and moving on until it falls into another. An *ergodic* system is one in which it is possible, in principle, to move from any state to any other in a finite time. A system that runs through a relatively simple cycle, where it passes through the same succession of states continuously, has a *limit cycle* or *periodic attractor*. A *strange attractor* is one that has the conventional properties of an attractor, in representing a relatively stable state, but which gives rise to non-periodic 'cycles'. A system drawn into a strange attractor will trace out a fractal trajectory rather than a simple cycle.

❑ *Behaviour.* The pattern of activity of a system over time.

❑ *Bifurcation point.* During the approach to a critical value of a control parameter a physical system can behave in a stable way. The bifurcation point is the point at which minutely different values of other variables can result in drastically different system behaviour.

❑ *Boundary.* The real or abstract delineation between a system and its environment.

❑ *Chaos.* The formal definition of chaos is stochastic behaviour occurring in a deterministic system. Stochastic means random or intrinsically unpredictable. Such behaviour occurs in deterministic systems which are extremely sensitive to small changes. This sensitivity can be so great that the long-term behaviour of a system can only be predicted (even in principle) given infinitely precise information about the system.

- *Collapse.* The uncontrolled decline of a system. This generally occurs when there are positive feedback effects eroding system limits, or when an event transgresses significant system thresholds.
- *Communication.* The transfer of information or, in social systems, meaning.
- *Connectivity.* The property of the structures that transmit effects through a system.
- *Control.* The process by which the system maintains its integrity or performance under changing demands.
- *Criticality.* A critical point is a point at which the behaviour of a system changes qualitatively.
- *Delay.* Time lag between cause and effect. Some elements of some systems take longer to react than others. This means that some events are synchronised, others are entrained but out of phase, while others are part of the same process of cause and effect but happening over different timescales. The degree of lag can itself be subject to threshold or interactive effects.
- *Equilibrium.* When a system is in a static or dynamic stable state. A system would achieve a particular dynamic equilibrium if, for example, the flows into a stock (see below) were the same as the flows out of a stock, so that the stock level was constant even though the contents of the stock were changing.
- *Emergence.* The phenomenon that systems have properties that the system components do not. An emergent property of a system is that which is not readily explainable with reference to subcomponents.
- *Environment.* Defined as that which lies outwith a system boundary.
- *Ergodicity.* An ergodic system is one in which it is possible, in principle, for the system to move from any state to any other state in a finite time.
- *Erosion.* A decline in the resource base supporting a system. This can occur inside a positive feedback loop, which means that the erosion itself makes a further erosion more likely to occur.
- *Exponential growth.* Growth by a constant fraction of the growing quantity during a constant time period, as with, for example, compound interest.
- *Feedback loop.* An iterating chain of causal connections. With negative feedback loops, change is effected in a direction that makes further change less likely. With positive feedback loops, change is effected in a direction that makes further change more likely. Where negative feedback loops tend to check and control growth, for example, positive feedback loops tend to amplify growth.
- *Flow.* Change of stock.
- *Hierarchy.* The effective structures, defined by levels of emergent properties, within which systems are constituted by and constitutive of other systems.
- *Hyper-region.* An area in a high-dimensional space.
- *Information.* That which reduces uncertainty. Technically, information is measured in the number of binary choices necessary to uniquely define an event. The human concept of meaning is rather more complex.

❑ *Input.* A flow into a system that is then transformed in some way.
❑ *Linearity.* A relationship that is proportional for all values of the cause and the effect and for which the effect of changing two or more control variables together is the sum of the effects of changing them independently. While some non-linear relationships can be approximated by linear models, in many complex systems, the non-linearities are both real and significant (see Chapter 3).
❑ *Model.* A descriptive intellectual construct.
❑ *Multi-factoriality.* The ability of a system component to serve both in a control function (causing change in another component) and in a dependent function (being changed by another component).
❑ *Non-linearity.* A relationship that is not strictly proportional for all values of the cause or the effect, or for which the combined effect of changing two or more control variables is not additive.
❑ *Output.* Flows from a system that have (usually) been transformed in some way.
❑ *Overshoot.* To go beyond the target value. This can be caused by delays in feedback, or an inadequate feedback process, so that the system does not adequately regulate. This can also be a function of rate of system change, as, for example, feedback delays that do not matter at slow rates of change may matter at rapid rates of change (see also Chapter 3).
❑ *Parameterisation.* The incorporation of small-scale phenomena and feedback loops into a model by taking average values for a range of effects.
❑ *Phase space.* A system can be represented as a point in a high dimensional space, termed phase space, whose axes are the control variables and whose coordinates are their current values.
❑ *Phase transition.* A phase transition, in a physical system, is a transition (see *Thresholds*) associated with simultaneous interactions on all length scales as when water freezes.
❑ *Resources.* Accessible sources and available means which can be utilised by a system.
❑ *Scenario.* An outline of what outcome to expect, given certain assumptions about starting conditions and system behaviour.
❑ *Sink.* A destination for a system flow.
❑ *Source.* A point of origin for a system flow.
❑ *Stability.* The ability of a system to resist perturbation.
❑ *State.* The state of a system is a complete description of every important aspect of the system at some time. In an unchanging system, the system state is unchanging, whereas in a dynamic system the state constantly changes as the system changes. In modelling systems we typically use equations which describe how one system state gives rise to another, and thus how the system changes over time.
❑ *State cycle.* A system's complete range of possible combinations.
❑ *Stock.* A store or quantity of material, energy, or information.
❑ *Structure.* The set of stocks, flows, loops and delays that define the

interconnections of a system. A system's structure determines the range of its behavioural possibilities. Although structures themselves do adapt, the term is normally used to refer to those elements that are permanent or adjust relatively slowly or infrequently.

❏ *System*. An interconnected set of elements, with coherent organisation. A system is characterised by hierarchical structure, emergent properties, communication, and control. Some systems can exhibit dynamic, adaptive, goal-seeking, self-preserving, or evolutionary behaviour. A *subsystem* is a component system, a *supersystem* is a superordinate system.

❏ *Thresholds*. Thresholds are points at which there is a qualitative change in the behaviour of an element of a system or the system itself. Threshold effects can appear for a number of reasons. For example, they can appear as a function of several independent constraints, where one constraint is inoperative within the bounds of the other constraint but operative outside of those bounds. Thresholds also appear in chaotic systems, where systems have zones of stable behaviour and zones of unstable behaviour. *Catastrophe theory*, which is the theory of sudden as opposed to continuous change, is sometimes used to describe what happens when systems collapse or radically transform at particular points.

❏ *Throughput*. The flow of energy, materials, or information from sources, via processing by the system, to the system sinks.

❏ *Transition*. A transition is a qualitative change in behaviour at certain critical values of control variables (see *Criticality* above).

3

Complex Adaptive Systems

A systems approach to sustainability entails considering the various agents interacting in the world as systems. Such an approach involves invoking general principles concerning systems to make inferences about likely and actual interactions between the systems under consideration. These principles can also be used to analyse the observed patterns of interactions between systems. Two particularly important and deeply interconnected phenomena which are common to the kinds of systems which are considered are evolution and adaptation, which form the focus of the present section.

Evolution and adaptation

The evolutionary biologist Richard Dawkins has characterised the fundamental agents of evolution as 'active germ-line replicators'.[20, 21] The idea is as follows. A *replicator* is any system which reproduces itself or which is reproduced. Such reproduction need not be perfectly error-free, because there is no reliably error-free reproductive process. While the most obvious examples of replicators are organisms and their genetic subcomponents, more prosaic items such as a piece of paper in the age of photocopying and faxing can also legitimately be considered to be replicators. Another interesting example is what Dawkins has termed a 'meme', which is a small unit of information. The observation here is that ideas reproduce themselves as people communicate them to one another.

Having established the idea of a replicator, Dawkins classifies these as *active* or *passive* according to whether the qualities of the replicator affect its likelihood of replication. So genes are active replicators because good genes tend to lend reproductive advantage to the corresponding organisms. This is because all 'good' genes ultimately improve either the organism's survival prospects or its likelihood of (successful) mating. This might, for example, be by making the organism more likely to avoid predation, better able to find food or more attractive to potential mates.

Similarly, memes are active, because ideas which are interesting or useful are more likely to be communicated than less interesting ideas. A piece of paper blindly photocopied might provide an example of a passive replicator, because its contents do not affect its chances of reproduction, but even in this case the reality is that pieces of paper with interesting or useful information on them are more likely to be replicated than less informative counterparts.

Having distinguished active from passive replicators, Dawkins then goes on to classify replicators as *germ-line* or *dead-end* according to whether or not they have the potential for being reproduced indefinitely often. Thus the non-sex cells in our bodies are dead-end, because even though they may reproduce a great number of times during an individual's life-time, once the individual dies it is certain that replication of such cells will cease. Sex cells, on the other hand, and more particularly the genetic material in them, are germ-line because they can be passed down through indefinitely many generations. Similarly memes and pieces of paper are germ-line, because there is no fixed limit to the number of times an idea may be passed on or a piece of paper copied.

These are profoundly important characterisations for the following reasons. Because no process of replication is completely error-free, a replicating system will inevitably undergo some variation as it reproduces or is reproduced. While it is the case that most errors introduced in the copying process are detrimental to the 'reproductive fitness' of the replicator (that is, make the replicator less likely to go on being reproduced) occasionally a random change will actually improve the chances of reproduction by enhancing the replicator. Of course, this can only happen if the replicator is active in the sense described above. If the replicator is germ-line, then there is the possibility of an indefinite accumulation of incremental improvement to the replicating system. This is what is meant by evolution, and this is the sense in which the term evolution is used here.

The idea of reproductive fitness is critical in the above. This is a measure of the ability of the replicating system to reproduce itself. Our fundamental understanding of natural evolution is based on the idea that any change which increases the number of viable offspring which an organism is likely to spawn will tend to propagate, whereas the rather more common changes which diminish such reproductive fitness will be disfavoured by evolution.

In the case of systems which reproduce themselves, there are two separate components of reproductive fitness. The first is the effectiveness of reproduction itself, which can – and indeed has – evolved. The second is any other factor which makes reproduction more likely, whether it be an enhanced ability to catch food, withstand harsh environments, attract partners or any of a myriad of other potential factors. It is also important to understand that the evolutionary fitness of an organism is not constant, but changes as evolution itself changes the environmental parameters and in the face of more general changes to the environment in which reproduction takes place. This results in an amazing variety of evolutionary effects. These include 'arms races', where, for example, a predator and prey co-evolve better mechanisms, the former for hunting and the latter for avoiding capture; symbiosis, which ranges from the co-evolution of cooperative behaviour or commensalist characteristics (like the African Oxpecker birds, for example, that pick insects from the ears of oxen, thereby providing themselves with a food source and the oxen with aural hygiene) to true symbiosis, which is where two species can no longer exist independently; and many others, including mimicry, camouflage and the evolution of intelligence, social behaviour and so forth.

Evolution is a special and very important case of a more general phenomenon seen in some systems, generally the more complex systems, termed adaptation.[22] Unlike evolution, general adaptation does not require the presence of active germ-line replicators. Examples include learning in humans and other animals, the development of muscles as a result of use, darkening of skin under the influence of sunlight, the biological processes that maintain an oxygen-rich atmospheric imbalance and the interlocking of stabilising feed-back loops discussed further on p 31.

Like evolution, adaptation can be explained in terms which require no higher organising principles, cognition or intentionality. They are in this sense 'blind' processes. The key consideration for an adaptive system is stability, rather than reproductive fitness. Stability plays an equivalent role, in this regard, to that of an adaptive advantage. Since no system is entirely isolated, variation is constantly introduced as the system is perturbed by outside influences. Moreover, many system states include dynamic processes which may have the effect of changing the system state. Certain system states will, however, have greater stability than others. This can be because they contain no internal dynamic for change, or because they include control mechanisms which limit dynamic behaviour to cyclic behaviour. It can also be because they have a greater tendency to resist change from external influences. Mechanisms for these sources of inertia include the development of negative feedback phenomena, which are discussed in greater detail on page 31, and bulk resistance phenomena.

The important observation about system states is that states with greater stability are, by that very virtue, displaced less quickly than their less stable counterparts. Similarly, subcomponents of systems which are more stable will inevitably tend to exist for longer. It is in this sense that in adaptive systems, stability replaces reproductive fitness.

There are a number of deep connections between adaptation and evolution. First, an ability to adapt can sometimes be evolved, as in the case of the human brain, which has evolved the ability to learn, and human muscles which have evolved the ability to grow. Moreover, complex adaptive systems may well contain evolving subsystems. The Earth system, for example, contains naturally-evolved stabilising mechanisms such as the Lotka–Voterra cycle, whereby numbers of predators and prey self-regulate. This happens simply because when there are too many predators for the current population of prey to sustain there is a famine among predators and many of them die, and when there is an excess of prey predators are able to breed more viable offspring, which in turn reduces the numbers of available prey.

Finally, it is possible to think about the 'evolution' of states in an adaptive system, in the sense of each state 'replicating' imperfectly to give rise to the next system state. In this unconventional sense, system states would qualify as active, germ-line replicators providing that the probability of replication was interpreted as a probability that the next state was similar to the present one. This is an unconventional perspective, since there is no population or differential reproduction in the usual sense, and such competition as occurs is for system time, with stable states being more successful at gaining system time than less stable states. In this context, however, the word competition is misleading because the present state has a monopoly on the finite resource of state space. For these reasons, we will use the more general term 'adaptation' to refer to general adaptive change, including evolutionary development, and we will reserve the term 'evolution' for systems which are genuine active germ-line replicators.

System thresholds and the global ecology

All biological and ecological systems have a degree of resilience. They will tolerate a certain level of stress or depredation, while maintaining the capacity to recover. Even if individual elements of a system are destroyed, these elements can often be restored provided that the essential network of relationships that constitutes the system remains. However, the expansion of human activities has led to the destruction of both elements of various systems and entire systems.

The global ecology maintains a dynamic homeostasis. This is achieved via factors that contribute inertia and negative feedback loops that will tend to compensate for and hence resist change.

There are a number of biological processes that play a role in maintaining the current ecological conditions. For example, biological activity is partly responsible for maintaining the amounts, proportions, and balances of gases that make up the Earth's atmosphere. The proportions of gases in the Earth's atmosphere are in a state of chemical disequilibrium, and are maintained in that state by biological feedback loops. If these biological processes were to cease to exist, the atmosphere of the planet would eventually become chemically stable, and would in turn cease to support most current forms of life.

Biological processes are dynamic, and respond to changing circumstances. For example, some four and a half billion years ago, when the solar system formed, the radiative temperature of the sun was about 70 per cent of today's value. Within the lifespan of the Earth, therefore, the output of heat from the sun has increased by some 30 per cent of its current value.[23] A 2 per cent decrease in the solar heat received by one hemisphere can be enough to establish an ice age. If the Earth's climate were determined solely by the output from the sun, the surface temperature of the planet would have been below 0°C until some 2 billion years ago, that is, for the first 1.5 billion years of the existence of life on Earth. The climate has, however, changed relatively little over this period. The planet was not frozen even when it received 30 per cent less heat from the sun than it does now.

The probable reason for this is that the atmosphere of the planet has always created a 'greenhouse effect'. This has varied with time, as during the course of its history the atmosphere of the planet has consisted of quite different balances of gases. For example, the pre-Cambrian atmosphere consisted of approximately 1.9 per cent nitrogen, 0 per cent oxygen, and 98 per cent carbon dioxide. The current atmosphere consists of 79 per cent nitrogen, 21 per cent oxygen, and 0.03 per cent carbon dioxide, with other traces. It was probably the atmospheric carbon dioxide, which can absorb terrestrial infra-red radiation and thereby delay its escape into space, that created the early greenhouse effect. Atmospheric ammonia may also have contributed to heat retention.[24]

Early forms of life would have increasingly taken up carbon, nitrogen, and hydrogen from the atmosphere. The carbon and the nitrogen would have been fixed (possibly deposited on the sea floor in the form of organic detritus), while the breaking-down of ammonia would have released nitrogen, leaving the hydrogen to form water and to escape as vapour. If there had been no other control factors, the depletion of the carbon in the atmosphere would have caused a drop in temperature, in turn causing ice and snow cover to increase. Ice and snow are highly reflective, so this would have resulted in more of the incident radiation from the sun being reflected back into space, which would have reduced the temperature to below 0°C.

There are several possible explanations for the fact that this did not happen. One is that sufficient methane may have been generated to provide compensatory heat retention. Another is that the Earth's albedo (the reflectance of the planet) may in fact have changed towards absorbing more rather than less heat. This could have happened if, for example, early micro-organisms became dark-coloured, so that large areas of the planet absorbed more heat.

It is clear from examples such as these that the process of biological evolution has itself generated changes in the ecological parameters of the planet. Such processes have been important in the creation of the conditions to which the human species is accustomed. For example, the evolution of organisms that excreted oxygen caused large reductions in the levels of atmospheric hydrogen and ammonia, and thereby made life possible for aerobic life forms.

The maintenance of a fine balance of oxygen in the atmosphere is now one of the biological processes of particular importance to humans. The atmosphere is currently 21 per cent oxygen. If the level rose to 25 per cent, the first spark would ignite even wet and cold timber and grass, and all combustible material in the world would burn. If the level fell much below the current balance, the productive potential of land-based forms of life, a significant proportion of which is now aerobic, would be significantly reduced.[24] Marine life would be relatively unaffected for some time, because of the reserves of dissolved oxygen in the ocean, but would be seriously disturbed if oxygen levels fell significantly below current values and the ocean started to become anoxic, which is what appears to have happened at the Permian–Triassic boundary event that eliminated some 95 per cent of all species.[6]

There are also positive feedback loops. These may involve just one factor. For example, if rising temperatures caused polar ice to melt, this would tend to reduce the reflectance of the planet, which would tend to cause a further temperature rise, and so on. More complex examples arise when two or more factors have a positive reciprocal relationship, so that, for example, growth in one causes growth in the other and vice-versa. Such relations can arise even when factors affect each other via intermediary factors, rather than directly. Positive feedback loops are sometimes enclosed in larger negative feedback loops, which maintain overall balance. Otherwise positive loops, once initiated, operate either until the positive feedback phenomenon saturates or until the system itself is transformed.

Humans interact with and affect a number of existing loops. The interaction with the greenhouse effect, via atmospheric pollution and the consequent additional global warming, illustrates a number of these points, and is discussed in detail on p 56. Whether the extent of human involvement in this particular set of cycles and loops will prove to be significant, either in terms of potential impact on human social and economic systems, or in terms of impact on global ecological systems, remains to be seen. However, as a general principle, rapid and extensive change in any system has a greater chance of transgressing a system threshold than slow and limited change. The rates of global warming, and the increases in the atmospheric carbon levels as a percentage of the total, would appear to be both rapid and extensive.

Ecological complexity and stability

In general, the more complex an ecology, and the more interlocking feedback systems there are, the more robust and better able to resist change the system appears to be. This is somewhat counter-intuitive, because there often appear to be more ways to break complex systems than simpler ones.[25] In practice, however, complex systems tend to exhibit greater stability. One reason for this might be that some complex systems, such as biological systems, employ multiple rather than single pathways of control. For example, human eating is governed by appetite, social convention, habit, time constraints and so on. These multiple causes tend to ensure the performance of the necessary act under a large variety of conditions. If eating were governed solely by hypoglycaemia, any disruption of the single control pathway would cause death. Systems that employ multiple determinacy may therefore tend to have enhanced chances of survival.[26] Similarly, genetic diversity appears to give an ecological system an ability to adapt to stress. One possibility is that the loss of a species in some human or other perturbation might matter less, in terms of the maintenance of a particular balance, if there are other species 'near' the ecological 'space' left by that loss that can then expand into that space. If no other such species exists, then the effects on the system itself may be more extensive. Another possibility is that a complex ecology that includes a number of species and a range of habitats may, by that very diversity and the associated uneven distribution of species and variants, be more likely to contain 'pockets' within which survive small populations of marginalised species or variants. Should the internal or external circumstances then change, and some dominant species decline, a previously marginalised species or variant may be able to expand. This could enable other elements of the ecological system to survive. The top predator in the system, for example, might be able to adapt its behaviour and possibly its shape to exploit this new food source.

More fundamentally, as is discussed on p 19, complex systems generally evolve from simpler systems rather than arriving in a fully-formed state. The incremental complexity will normally develop only if it confers a net adaptive or selective advantage to the 'parent' system. Part of this advantage may accrue precisely because the complexity makes the system more robust. Even if this is not the case, the more complex system will only be able to survive if it is highly robust, because it will inevitably suffer many interferences during the course of its existence. This is true to a greater extent for complex systems than simple systems, because there are more parts to be interfered with, so stability is a natural characteristic of these systems. This relates closely to the weak 'anthropic' principle in physics.[27]

The global ecology is a very complex system. It is clearly robust, and will tolerate a degree of stress. However, complex systems almost invariably have thresholds. Once such systems are pushed beyond critical thresholds, which can happen when, for example, vital negative feedback loops are broken, they typically undergo some sort of transition into a new state. The speed of the transition will depend on the nature of the system and on the precise circumstances of the change, as will the relative stability of the new state compared with the one that obtained earlier. This 'punctuated equilibrium' behaviour can be seen in biological evolution, for example, where forms and relationships that have been stable for millenia can transform relatively rapidly when populations are under stress. There is also increasing evidence that the climate can transform relatively rapidly in certain circumstances (see page 59).

Like other species, humans require from the global ecology both a high degree of stability and a particular kind of dynamic equilibrium that provides certain essentials, such as the right amount of oxygen and so on. At any one time, the global ecology will be in a highly unusual state in providing conditions suitable for all the systems currently in existence, and therefore any change away from the current ecology is always likely to be detrimental to a number of species and systems.

The global ecology must have thresholds. It is therefore possible that it could undergo a transition into a state which would not favour humans. Such a transition might occur because of purely external factors, and it is unlikely that much can be done to avoid or reduce the likelihood of such a possibility. It is possible, however, to ask whether current or foreseeable patterns of human behaviour are themselves likely to increase significantly the probability of changes to the global ecology which would make human life more difficult or impossible.

There can be no certainty in such predictions, partly because the complexity of the global ecology exceeds current understanding, and is likely to continue to do so for some time. Another cause of uncertainty is the possibility that some sub-systems of the global ecology may, under some circumstances, manifest chaotic behaviour, which is effectively indeterminate, and which would make exact predictions impossible. The combination of partial knowledge and understanding, very high complexity and non-linearity, adaptive and evolutionary phenomena and chaotic indeterminism ensures that prediction of outcomes can only be probabilistic, and the degree of accuracy may be low.

Despite these limitations, the following observations can be made:

❑ Regional ecologies and the global ecology must contain thresholds, beyond which they would transform to new points of dynamic balance.

❑ It is therefore possible that regional ecologies or even the global ecology could undergo a transition into a state less favourable for humans.

❑ There have already been many examples of human actions which have resulted in regional ecological change which have made human life more difficult or impossible, such as nuclear and chemical toxification of areas, desertification, acidification, salination and so on.

❑ It is not known for certain whether any human action could induce a transition in the global ecology which would make human life impossible, or how likely it is that humans could have such an effect.

❑ Some ecologies become less stable when impoverished or otherwise significantly altered than in their antecedent state. Ecologies altered to the point where they enter an unstable phase will tend to carry on changing until they reach a new equilibrium. Reaching this equilibrium may involve further losses.

❑ It is clear that human actions are causing changes to ecosystems and other systems in the biosphere, the troposphere and the stratosphere. Some of these changes are relatively large, and some are occurring at rates that make adaptive and evolutionary response very difficult. It is possible that no combination of changes of this magnitude has occurred since the major extinction boundaries.

❑ If the levels of environmental impact, including the reduction in genetic diversity, continue at current rates, the likelihood of regional and possibly even global ecological instability must tend to increase.

This suggests a policy of prudence, and an attempt to stabilise or reduce aggregate environmental impact and rates of change at least until such time as knowledge of the behaviour of complex adaptive systems in general and the global ecology in particular makes it possible to estimate the likely location of system thresholds with a degree of confidence.

Risk

All models of the behaviour of complex systems, such as environmental or economic systems or the interaction between these two, are imprecise and uncertain or limited in their scope. These uncertainties and limitations arise for a number of reasons. One is that knowledge of these particular systems is incomplete. Another is that knowledge of the behaviour of complex systems per se is incomplete. A third is that the behaviour of some complex systems is, at least in some areas, indeterminate. This means that the behaviour of such systems can only be modelled in probabilistic terms, in limited domains, or a finite time into the future. All of these considerations probably apply to the complex of complex systems, the environmental, social, and economic systems, that make up the world in which we live.

This means that it is usually necessary to make some estimate of the error and risk margins associated with every possible predicted outcome.

This gives rise to a number of issues. The calculation of risk is itself accompanied by a degree of error. Such margins can be large, and can therefore determine the outcome of decisions. Furthermore, each possible outcome may be attended by a different distribution of risk, that is, by a different distribution of costs and benefits. This means that there are technical and political considerations in any analysis of probability and risk. Consider the following example. A small risk of a major disaster (such as a 0.01 per cent chance of an incident that would kill 10,000 people), and a larger risk of a smaller disaster (such as a 100 per cent chance of an accident killing one person), give rise to the same expected outcome in terms of losses (1 statistical life), provided that the risk estimates are accurate. Despite this, the characteristics of the two kinds of risk are very different, and people are sensibly concerned about the distribution of such risks. There is no 'correct' way of choosing between such risks, but it is clearly important not simply to aggregate the information in such a way as to disguise the difference between the two.

Furthermore, it may in real instances be necessary to compare the kind of risks above with a practical certainty of minor effects on much larger numbers of people. This makes it necessary to assign priorities to the interests of the people concerned and affected.

Perhaps even more difficult than determining which course of action is more appropriate, given a full and accurate picture of the risks associated with different courses of action, is the initial assessment of risk. For example, the likelihood of nuclear power station failure is often assessed by accumulating the probability of each component failing. In practice, operating experience suggests that the relatively unlikely event sequences assessed are not typical of real system failures. Actual failures are more often associated with operator errors of kinds that are not anticipated and are therefore not included in the risk assessment. If risk is computed by statistically analysing the track record of existing installations, very different assessments are generated.

There are thus serious technical problems with assessing the risk even of those human-built systems which it is generally assumed are well-understood. In the case of complex natural systems, which are generally less well understood, risk assessment becomes even more problematical. Given current knowledge and the other factors referred to above, it is often not possible to determine when a system under some form of stress is approaching a transition threshold. It is very important to note that the 'straw that breaks the camel's back' is not a special straw. As such a point is approached, the cost of continuing with a given pattern of behaviour rises sharply, but this may not be apparent until the transition threshold has been passed. This example illustrates the non-linear nature of risk. Non-linear phenomena are discussed in general below.

Perceptions of risk

It is also important to note that perceptions of risk will not necessarily reflect the actual risks involved in a situation. Most people will rate a situation as being significantly more risky if, for example, their exposure to the risk is not voluntary, if children are exposed to the risk, or if the situation is not perceived as fair, that is, if the benefits accrue to one person and the risks to another.

Thus many North Americans rated their exposure to the agrochemical Alar, which was sprayed on apples, as significantly more risky than drinking high-roast coffee, even though both contain potential carcinogens and the latter activity is arguably the more dangerous. Exposure to Alar was widely perceived as a very high risk partly because children were exposed to the risk, because the exposure was not undertaken knowingly or voluntarily, and because the benefits of spraying accrued to the producers and retailers while the consumers accrued the risks. In this case, the apples were slightly cheaper as a result of the use of Alar, but this was not seen as adequate compensation by the consumers concerned.

Such social and psychological factors are highly important considerations when translating technical assessments of risk into the terms of everyday language and experience, and when formulating procedures for controlling risks in the domain of public policy.

Non-linearity, criticality, thresholds, and transitions

The distinction between linear and non-linear systems is fundamental. Suppose that some behaviour measured by a variable y is controlled by some factor which can be quantified as a variable x. For example, y might be the temperature of the water in a kettle, and x the amount of electrical energy supplied from the mains. Then y is said to be a function of the control variable x. This allows graphs to be drawn showing how y changes as x is changed. Examples are given in Figures 1 and 2. Suppose that

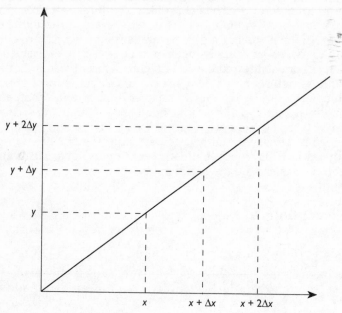

Figure 1: A dependent variable y is said to be a linear function of a control variable x if the graph of y against x is a straight line, as shown

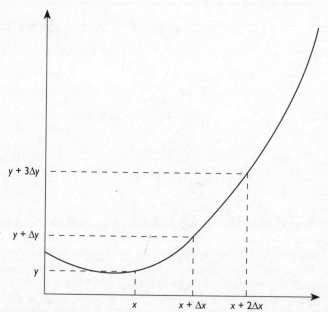

Figure 2: A dependent variable y is said to be a non-linear function of a control variable x if the graph of y against x is not a straight line, as shown

changing the control variable by an amount Δx results in a change in the dependent variable y by the amount Δy, as shown in the figures. The relationship between x and y is said to be linear if changing the control variable by twice the amount (by $2\Delta x$) always results in twice as large a change in the dependent variable y (ie it will change by $2\Delta y$). In the present example, this would mean that doubling the electrical energy input would double the temperature rise, which is approximately true for an electric kettle within certain temperature ranges. Linear relationships between a pair of variables are characterised by straight-line graphs, such as that shown in Figure 1, hence the name.

In the more general case, y may depend not just on a single control variable x, but on a number of control variables $x_1, x_2, ..., x_n$. For example, the temperature of the water in a kettle also depends on the pressure, which changes with altitude. This more general relationship between y and all the control variables x_i is said to be linear if not only is y linear in each of the control variables x_i (so that the graph showing how y changes with each x_i while fixing all the other control variables is a straight line), but also the effect of changing two or more variables together is the sum of the effects of changing them each individually. Thus if temperature were linear in energy input and pressure this would mean that if adding an amount ΔE of energy raised the temperature by 10°C, and increasing the pressure by Δp raised also the temperature by 10°C, then the effect of adding ΔE of energy and raising the pressure by Δp would be to raise the temperature by 20°C.

Thus the two essential characteristics of linear systems are that the variation of the dependent variable on each of its control variables has a straight-line graph, and that the effect of each control variable is independent of the effects of any of the others. Linear systems are very much simpler to analyse than non-linear systems, and they are much better understood. The most important and widely used tools for analysing system dynamics are differential equations, which relate rates of change in various variables. Until recently, scientists have confined themselves almost exclusively to studying linear differential equations, and linear systems. When linear differential equations have not been available, the standard approach has been to look only at small changes in control variables and to linearise the true system of equations by discarding the non-linear parts of the description of the system.[28] This has led to a rather distorted and simplified view of the world. Linearisation not only reduces the accuracy of the predictions of the equations and limits their application to small perturbations of a system, but can also be significantly misleading in that in many cases the qualitative behaviour of the linearised system is also different. In particular, linear systems will not exhibit chaos, and will tend not to exhibit the same sort of critical behaviour as is seen in true non-linear systems.

In natural systems, linearity is the exception rather than the norm. It is true that there are often approximately linear relationships, but these tend to occur within bounded domains. That means that when such a system is taken beyond a certain point, its behaviour will change.

A typical example of this is seen in many growth phenomena. Many systems, both natural and artificial, trace out 'S'-curves such as that shown

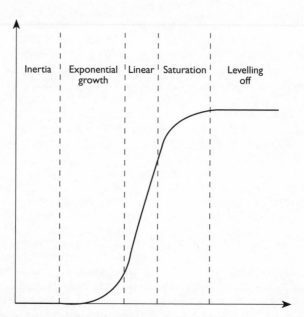

Figure 3: Many growth phenomena, both natural and artificial, follow patterns which conform broadly to the sort of 'S'-curve shown above

in Figure 3. This is quite easy to understand in the following way. Most systems respond to change slowly at first, because if they did not, they would tend to be unstable. However, a typical pattern of response once the 'inertia' of the system has been overcome is for the system to experience an exponential growth phase, before passing through a phase in which change is approximately linear. This linear phase is sometimes relatively short-lived. As this is a finite world, there is a limit to most measurables. This inevitably means that the change eventually levels off, or a further transition occurs. The levelling off may be caused by a limit to the available input for the process (whether materials, land-space, collectable energy or whatever), the onset of another process which counters the growth phenomenon or simple saturation. The net effect of the factors outlined above is a characteristic S-curve. It is sometimes useful to think of S-curves as being composed of three approximately linear sections, joined by curves.

Another important concept is that of criticality. A brick resting on a plank will start to slide when the plank is tilted up beyond a certain point. However, until the angle increases beyond some critical value (known as a critical point), friction suffices to counteract gravity. At that critical point the brick begins to move, radically changing the qualitative behaviour of the brick. At that stage, provided the critical angle is small, acceleration is approximately proportional to the difference between the angle of the plank and the critical angle, and is therefore linear.

In physical systems, phase transitions occur at critical points. At such a point there is typically a significant qualitative change in a system's behaviour. Water, for example, has three basic phases: solid (ice), liquid

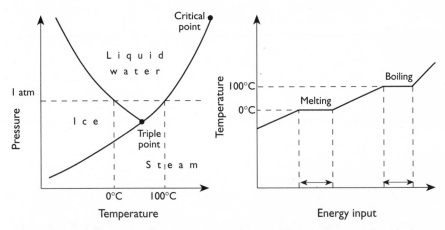

Figure 4: The diagram on the left is a simplified 'phase diagram' for water, showing the behaviour of water at different pressures and temperatures. The lines show the boundaries between water's different 'phases' – ice, liquid water and water vapour (steam). Phase transitions occur at the critical values on the phase boundaries and, in the case of water, energy must be put into or extracted from the system to effect a phase transition. This is illustrated in the graph on the right

(water) and gas (steam)(see Figure 4). Critical points occur (at normal atmospheric pressure) at 0°C (the freezing/melting point) and at 100°C (the vaporisation/condensation point). On the approach to these critical points temperature change is approximately linear as energy is put into or taken out of the system. At the critical points different phases co-exist, and it takes energy to move past these particular phase transitions (at least, at typical pressures).

Phase transitions in physical systems have a precise technical definition, and are associated with simultaneous interactions on all length scales. The notion of a qualitative change in behaviour (a transition) at certain critical values of control variables is more widely applicable, so we will use the term 'transition' in this sense.

A particularly important kind of critical point is a bifurcation point. Here, during the approach to the critical value of the control parameter the system behaves in a characteristic (and usually stable) way. Beyond the bifurcation point, minutely different values of other variables result in drastically different system behaviours. There are frequently sequences of bifurcation points, each behaviour splitting into two sub-behaviours at each bifurcation point, often in an infinite cascade, leading to the notion of 'deterministic chaos'. The most notable feature of these bifurcation phenomena is that they arise even in some very simple systems.

Biological organisms often exhibit critical behaviour. This may be very simple. For example, as the level of some particular environmental stress (such as a toxin) is increased, organisms will generally survive up to a given level of toxicity, but beyond some critical point will die. In some cases, the critical values of interacting systems must coincide. In the course of the development of the human embryo, for example, the embryo is susceptible to given levels of environmental toxins to different degrees at different stages of development.

The speed of change, especially those changes that incur a transition, is usually of considerable practical significance. Species, for example, generally evolve over a time scale measured in many generations. Such evolutionary time scales are, characteristically, millions of years long.

Species become well-adapted to their current eco-niche, and can usually survive many of the normal small perturbations to the conditions within that eco-niche. Adapting to larger scale changes is usually much more problematical and is not always possible. This is for a number of reasons. One is that a larger change is more likely to incur a transition. Another is that systems tend to have varying abilities to resist change in different dimensions. In particular, systems may have relatively little ability to resist change of a type to which they have not previously been exposed. It is also possible for a system to lose the ability to resist some environmental pressures should such pressures be removed for a sustained period, as the process of evolution tends to exert a mild pressure to discard parts of a system which no longer make a useful contribution if there is any 'cost' associated with carrying the redundant part.

For example, if the ozone layer were to thin slowly over many generations, allowing the levels of UV-B radiation reaching the surface of the planet to increase very gradually, humans might slowly become more adapted to cope with such radiation. They might become more resistant to the consequent skin cancers and eye cataracts. With the currently observed rates of ozone depletion, however, such an evolutionary response is probably impossible. This is likely, therefore, to require changes in living patterns in order to avoid sunlight, or the use of some 'technofix' such as wide-brimmed hats, dark sunglasses, and effective UV-blocking cream. There is also a possibility that the modern health care now available in varying degrees will make biological evolution of the kind described less likely to occur, as the prospects of survival and hence reproduction of the more vulnerable members of the species may be enhanced by medical or surgical intervention.

A systems model of sustainability

The mathematical models of system dynamics can be used to integrate the principles discussed in this section into a model of the behaviour of the highly complex system that constitutes the planet.

In this integrated model, the central problem of sustainability can be stated concisely in terms of *phase space.* The very complex system that constitutes the planet, which contains a number of interacting complex sub-systems such as biological, ecological, social and economic systems and so on, can be represented at any time as a point in a high dimensional phase space whose axes are the control variables and whose coordinates are their current values (see Figure 5). Technically, every independent control

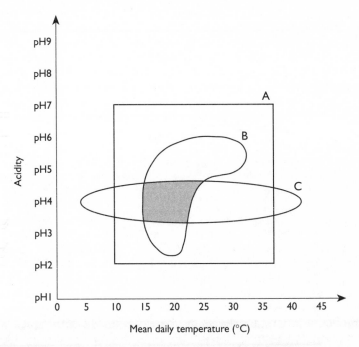

Figure 5: A grossly-simplified two-dimensional section through phase space for the Earth. The regions shown on it suggest possible survival regions for three systems, A, B and C. System A can itself tolerate a wide range of acidities and temperatures, but it depends on systems B and C, which cannot. Thus the effective survival region for system A is the intersection of those for A, B and C, the shaded area shown

variable is included as an axis in phase space, so that the state of the system is entirely defined by the point in phase space at which the system resides, as is its full history and future (subject to true indeterminism).

For certain of these control variable axes there is a small subset of values that humans can tolerate. More precisely, there is a region (or hyper-region) within which the entire range of systems on which humans are dependent can survive. In other words, this region is the intersection of the sustainable or survival regions of all systems necessary to continued human existence. In general, as the boundaries of this region are approached, human existence will become more difficult and dangerous, and less comfortable, as humans and the systems on which they depend come under increasing stress and approach major transitions.

If, in the course of time, the planet system moves to a point in phase space outwith the tolerable limit for one or more of the systems on which humans depend then the human species will not survive. Human survival depends on the system remaining within the small subset of all possible outcomes in which it is positioned, within the tolerable limits on all of the critical control axes. This region in phase space defines sustainability (see

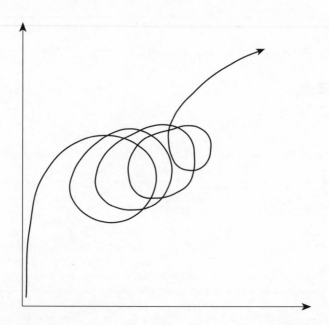

Figure 6: This diagram shows the trajectory of a hypothetical system through a two-dimensional section of phase space. While many approximate cycles may occur, it is unlikely that the trajectory can ever intersect itself (in all dimensions)

Figure 5). It will not be symmetric, as humans (as with most species) are able to adjust to wider ranges of tolerance in some dimensions than in others. Moreover, it is unlikely that all of systems on which human existence depends are currently known.

The full phase space for the planet is clearly of a very high dimension. Some of the axes are likely to be related. This means that, in purely formal terms, some of the axes may be redundant. However, there may be practical reasons (such as ease of measurement) for using some technically redundant axes to assess movement towards or away from sustainability in an approximation of phase space. Of course, some positions in phase space cannot be reached. As a trivial example, the relative levels of nitrogen and oxygen in the atmosphere may alter, but it would be impossible for the atmosphere to become 100 per cent nitrogen and 100 per cent oxygen at the same time.

The Earth moves through phase space over time, tracing out an arc. This movement contains many approximate cycles, and has many different characteristic time scales (see Figure 6). It is unlikely that the system can ever return to any given point. This is because the position in phase space completely determines the state of the system, so that if a system were ever to return to some previous point it would be bound to cycle forever, subject only to true indeterminacy in the universe.

The Earth system is in some respects similar to biological systems. The most important difference is that the Earth system does not replicate.

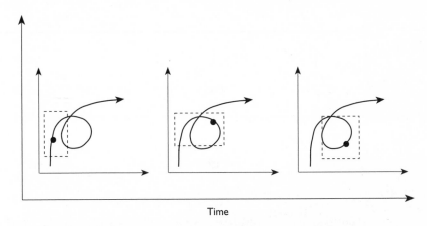

Figure 7: As the Earth moves through phase space the survival region of humans will tend to follow the arc traced out, changing shape in the process. This will happen as a result of evolutionary and adaptive change, and possibly also as a result of the development of technology. There is, however, no guarantee that the survival region will, in fact, be able to track this arc successfully, and the faster the movement, the more difficult and unlikely will this tracking be

However, it is similarly subject to evolution and adaptation, although the mechanisms are different. More precisely, system states evolve as the Earth system travels through phase space (see Figure 7).

There is a general tendency for all such systems to move towards stable states. This is simply because the system will by definition spend more time in stable states. Of course, such stability is relative. No state is ever completely stable, and although the system will 'lock into' regions of stability, it will always remain in motion.

This is a general principle of system behaviour, and can be seen in the pattern of evolution of life, the way in which ecological and biological systems maintain stability and adapt, the way in which economic and social systems evolve, and so on.

As the Earth system changes, the environmental factors that shape the outcome of the evolutionary process also change. This alters relative selective pressures and the balance of evolutionary advantage, and adaptive systems will naturally respond to this. This means that the region around the current position in phase space that defines sustainability itself changes and, with a delay, follows the current point.

Of course, there are critical points beyond which any given system will not be able to go, and when these are passed, the system concerned becomes extinct or transformed. This means that irreversible losses will naturally occur in the course of the planet's movement in phase space. Current and future irreversibilities will affect the human species to varying degrees. The continuing survival of the human species depends on our avoiding an irreversible state transition on a critical control variable axis.

The speed with which change happens is also important. The delay factor with which the phase space region that defines sustainability moves to follow the current point at which the planet is positioned in phase space will, for biological systems, be measured in terms of the evolutionary timescale, that is, in terms of large numbers of generations. This means that, for any individual human or generation of humans, the region in phase space that defines sustainability is normally relatively fixed, though technology can make otherwise inaccessible regions of phase space habitable at the cost of an indefinite commitment to supplying this technology.

The interaction between humans and their environment can be thought of as moving the planetary system along various axes simultaneously. If this happens at a rate that exceeds the rate at which other systems can adapt, that is, at a rate that exceeds the delay factor with which the phase space region that defines sustainability can move to follow the point at which the planet is positioned in phase space, then these other systems will become extinct. The species extinction rate (see p 72) therefore provides a partial measure of the rate of movement through phase space.

Global warming (see p 56) is likely to provide a number of examples of this effect. Vegetation distribution typically shifts some 200 km towards the poles with each 1°C rise in temperature. The forest migration rates at the end of the last glacial period were some 20 to 100 km per century. However, the projected rate of global warming will be perhaps 100 times faster than the rate of warming at that time. Many tree species will be unable to migrate at the necessary speed, which will in turn affect a large number of dependent species.

The general implications for policy

The key question in the analysis and development of the political and economic policy options for a transition to a more sustainable way of life is that of flexibility, where flexibility is defined in terms of uncommitted potentiality for change.[26] If a critical control variable is at an upper or lower bound the loss of flexibility is likely to be pervasive. The total range of possible options tends to become increasingly rapidly circumscribed as a critical control variable approaches such a boundary. In such a situation, it may be necessary to prioritise those changes that alter the position of such a control variable to bring it back within tolerance bounds, regardless of other costs or distortions. Such ad hoc responses can lead to a cascade of other adverse consequences.

Take a transport problem as a simple example of how options become constrained. If car numbers and usage were increasing, and if roads were to become congested to the point where traffic could not move, and if restraints on car use were ideologically unacceptable, then an ad hoc response might be to build more roads. This would provide a temporary relief until the newly extended road system re-congested. However, an increasing cost would be incurred in terms of the further transfer of resources from other modes of transport, which would increase pressure to abandon these other modes of transport, and so increase the road congestion. In this way, flexibility is rapidly absorbed or consumed.

Human population growth can also be thought of as a way of absorbing flexibility, especially in conjunction with a refusal to accept certain 'compensating' factors, such as famine and disease.

Any solution to the problem of sustainability therefore has to solve two connected problems:

❑ Flexibility must be found or created.
❑ The system must be prevented from immediately absorbing the new flexibility.

Translating between systems

The need to solve these questions clearly has a number of implications for economic systems, which are discussed later. Before going on to review these implications, it is important to consider how economic and environmental systems and theories relate.

To summarise some of the points in the previous chapter, different modes of behaviour are manifest at various organisational levels. For example, qualitatively different phenomena seen as particles are grouped into atoms, atoms into molecules, molecules into cells, cells into organs, organs into organisms, organisms into social groups, and so on. It is important to realise, however, that these groupings are essentially analytical constructs which help us to classify and understand complex phenomena when studied at different temporal and spatial scales. There are deep structural connections between these organisational levels, and in principle it should be possible to explain all higher-level phenomena in terms of more basic levels. It is, however, often inappropriate, difficult, or practically impossible to explain some behaviour at a given level of reality in purely reductionist terms – that is, solely in terms of the aggregated behaviour of the constituent parts. Partly for this reason, as far as scientific theories and models are concerned, levels of reality are not always reducible to lower levels. As a general principle, all theories in natural and social science are approximations to reality. The existence of boundaries means that scientific theories have a certain domain of validity. Beyond this domain, each theory will no longer give a satisfactory description.

In reality, many if not all processes and systems are multifunctional. For example, a single organism will often be both predator and prey in different contexts. The global ecology is composed of many arbitrarily bounded complex systems, which interlock and exhibit symbiosis, parasitism, competition, cooperation and so forth. This inevitably means that aspects of systems intersect and project into more than one domain, often including the domain conventionally identified as that of economics. It is only the elements which directly impinge upon the economic sphere that a purely economic theory normally attempts to capture. However, wider analyses can then integrate a range of theories and disciplines that bear on a common subject, drawing in higher-order phenomena and interactions.

As theories develop, they are either replaced or extended to give an improved approximation, or integrated with other partially successful models. In a general systems theory approach, a scientific critique is achieved by checking models for self-consistency and compatibility with other existing models.

Each domain, such as that of economic or of environmental systems, is a subsystem of a larger system. For most purposes, these hierarchies culminate in the Earth system. Subsystems vary in the extent to which they are open or closed systems. However, no real systems are completely closed. Subsystems such as economic systems are very open systems, in that they interact with other subsystems in a great many ways.

This becomes especially important when developing policy to regulate a pattern of interactions between open systems, such as economic and environmental systems. The theories of subsystem behaviour have to be sufficiently congruent. This does not mean that theories of the behaviour of one subsystem can be adequately subsumed in the theories of the behaviour of the other subsystem. It means that both subsystems must be adequately known and understood, that the relevant inputs and outputs from both systems must be described in similar terms and, most fundamentally, that the respective theories of subsystem behaviour reflect the open nature of the systems and can therefore encompass the patterns of mutual interactions.

Given the current lack of knowledge of the relevant system behaviours, attempts to incorporate environmental factors into economic decision-making by monetising environmental variables, and thereby mapping one system into the values of another, are unlikely to contribute to a real understanding of the system dynamics, although such techniques have an important role to play in clarifying policy options and in effecting political and economic change.

4

Environmental Factors

Chapter 3 ended by suggesting that an understanding of sustainability requires the integration of information from a wide range of disciplines, because sustainability concerns the interplay between systems, such as economic and environmental systems, which are usually studied by separate disciplines.

This chapter reviews some of the environmental dimensions of the issue. These have been selected on the basis that there is some accord that they indicate particularly large-scale and significant changes in the global environment, some of which may even, if unchecked, induce significant and potentially uncontrollable changes in the current global ecological balance, or because they indicate possible threats to human social and economic systems, or imply restrictions on the future pattern of human interaction with the global ecology.

It should be noted that these indicators may not be of equal salience. This is for a number of reasons. One is simply that some have received more attention recently than others. Another is that while people may be aware of environmental damage or loss that occurs within their lifetime, they generally accept as the norm the environment that they inherit. They may therefore be less aware of those issues where there has been an accumulation of environmental damage caused by previous generations, such as with the gradual deforestation of the UK.

It is also important to note that many of the figures relating to the extent of various environmental changes and impacts are relatively uncertain. The literature contains a fairly wide range of estimates for some of the more obvious environmental changes. There have been particularly contentious discussions as to the extent of desertification and global warming, for example. This uncertainty reflects a number of factors.

❑ Many large global systems – such as the biological dynamics of the oceans – are not, as yet, well known or understood, although this is not to underestimate the extent of the knowledge that we do have of some of the physical and biological systems.

❑ A great deal of relevant data has only recently been collected, partly in response to the growing interest and concern about environmental change. Satisfactory data is still lacking, in many cases, or has not been collected for long enough to be able to extrapolate a trend with a reasonable degree of certainty. It is often necessary to qualify projections with relatively large margins of error.

❏ As part of the normal process of scientific change, it is also sometimes necessary to revise older theories as well as estimates, which means that there must always be a degree of doubt that the current models for interpreting the available data are fully adequate.

❏ Many estimates involve making judgements on the basis of incomplete data. Such judgements are prone to various forms of conscious and unconscious bias from a number of sources, such as the prevailing scientific paradigm, peer group pressure, and the need to produce results likely to lead to further funding for research.

❏ The extent of obvious environmental impact and change varies greatly between different regions. Scientists must, as a rule, look in detail at a sample of sites, then extrapolate in order to calculate the total extent of wider change. The choice of sample sites can, therefore, have a significant effect on the final estimate. Over time, of course, more observations can be made at more sites, and estimates improve.

In this chapter we report a range of estimates of environmental change, all fairly recent, but it is important to bear the above caveats in mind. Most projections, regardless of whether they indicate catastrophe or utopia, should be regarded with some caution. This uncertainty gives rise to a number of issues, which were reviewed in the last chapter.

The important general points to bear in mind, however, are as follows:

❏ All visible environmental effects are indicators of change, which may be more (or less) extensive than is immediately apparent.

❏ It is the compound scale of the effects that may prove to be significant, rather than any one of the dimensions of environmental change, as this indicates both the extent of the human ability to alter the conditions for life on Earth and the potential for unanticipated synergistic interactions between different dimensions of the problems, which could lead to their rapidly becoming more intractable.

❏ While change per se is inevitable, all change is dangerous for somebody or something. It is not necessarily the main agent of a change who will be placed at risk by that change. It is possible for the risk or the damage to accumulate elsewhere.

❏ Large changes will always tend to be more risky than small changes.

❏ While all but one of the indicators of environmental change may prove to be unimportant, it may only need one significant environmental change to cause immense disruption. With reference to the communique issued by the Irish Republican Army after their bombing of the Conservative Party Conference in Brighton, in which they made the point that they only had to get lucky once, while the security services had to be lucky all the time, we have termed this the 'Terrorist Principle'.

The biosphere

The surface area of the planet is 510 million km^2. This is distributed as shown in Table 1 (please note that the figures in this and in all other tables are rounded), which shows that nearly three-quarters of the surface of the Earth is under water.

Table 1: The planet's surface[29]

Area type	Area (M km^2)	Per cent of land	Per cent of total
Uninhabited land			
Mountains	9	6.0	1.7
Tundra	12	8.0	2.3
Forest (mostly rainforest)	14	9.4	2.8
Glaciers	15	10.0	3.0
Deserts	40	26.6	7.8
Uninhabited land total	*90*		*17.6*
Inhabited land			
Arable	34	22.7	6.6
Other (cities, industrial etc)	26	17.3	5.2
Inhabited land total	*60*		*11.8*
All land surface total	150	100.0	29.4
Oceans	360		70.6
Surface of planet total	*510*		*100.0*

The floors of the deepest oceanic trenches lie just 11km below the surface of the sea (the average depth of the ocean is 3.7km), while the highest point on Earth, the summit of Mount Everest, is less than 9km above sea level. Thus life exists in a zone only 20km deep, a distance that an athlete could run in little more than an hour, more than half of which is under water. The inhabitable part of the Earth, therefore, is like a thin film over the surface of the planet.

Oceans: resources and pollution

There is one continuous ocean on the planet, partially divided by land masses. The main subdivisions are into the North and South Pacific Ocean, the North and South Atlantic Ocean, the Indian, the Arctic, and the Southern Oceans, and a number of seas.

Every land mass has a continental shelf, which is actually submerged coastal land. These vary from 0 km to 1,500 km wide, averaging about 70 km wide. The exposed width of the continental shelves depends on sea level. During the late Cretaceous period, some 65 million years ago, the sea covered 82 per cent rather than 71 per cent of the Earth's surface. During the glacial periods of the Pleistocene, from 2 million to 10,000 years ago, the sea level would fall due to the accumulation of ice on land. At the peak of the last ice age, sea level was 130 m below the current level, so most continental shelves were exposed. During interglacial periods, such as the one in which we live today, the sea level rises and most continental shelves are submerged under some 100 m of water. Britain was cut off from the rest of Europe about 8,000 years ago by the rising sea level. The true ocean basins start at the outer edge of the continental shelves, and are mostly 4 km to 6 km deep.[30]

There are a number of economic uses of the ocean:

❑ The principal non-extractive uses are as a means of transport, especially of freight, and as a waste sink.
❑ The principal extractive uses are biological harvests, mining for minerals, and utilisation as an energy source. Desalination and conversion to fresh water is of critical importance in some areas, but is not significant globally.

Non-extractive uses

Shipping

There are about 430 million gross registered tonnes of commercial shipping (1983 figures). About 80 per cent of world sea freight consists of petroleum, chemicals, and bulk dry cargoes such as iron ore, coal, bauxite, phosphate, and grain. About 1 million tonnes of shipping is lost each year, which can add to the problems mentioned in the next section.

Waste disposal

Water systems (see p 54) are used, often simultaneously, as a source of water for drinking, irrigation, and industry, and as a means of disposing of sewage, and agricultural and industrial waste. As countries develop, water usage shifts from agriculture to industry, the type and volumes of pollutants change accordingly.

The ocean can absorb significant but probably not unlimited flows of human waste. Impacts of waste flows have been small globally, but high in some waters, such as estuaries, coastal regions, and semi-enclosed seas like the Mediterranean and North Seas.

The main categories of waste are:

❑ sewage;

❑ industrial wastes;
❑ dredgings from rivers and harbours;
❑ spoil from land works; and
❑ thermal waste.

Flows of waste can be continuous or episodic, deliberate or accidental. For example, some 2 to 5 million tonnes of oil reaches the ocean each year, of which about 95 per cent is deliberately discharged and about 5 per cent is accidentally discharged. Much publicity is attached to accidental discharges, which tends to give a misleading impression of their contribution.

The oceans have vertical circulation periods of 200 to 1,000 years. This determines the length of time water stays out of contact with the surface. This in turn determines the safety margin for disposal of waste in deep water, as certain wastes, such as some forms of nuclear waste, may remain dangerous for longer than the circulation period.

A semi-enclosed sea will have a certain water change rate. The flushing time of the North Sea, for example, is about a year. This limits the build-up of local pollution by distributing it, eventually, around the world ocean.

Extractive uses

Biological harvest

The world annual total marine harvest reached some 70 million tonnes of fish and 3 million tonnes of plants (with smaller quantities of shellfish and marine mammals) in 1986, and has continued rising. There are 11 countries, including the UK, that harvest 0.5 to 1 million tonnes, and 20 that harvest more than 1 million tonnes. The largest consuming nation is Japan, which takes about 12 million tonnes, and the former Soviet nations, which take about 11 million tonnes. Fisheries provide about 50 per cent of human protein intake in Japan, and 20 per cent in the former Soviet nations, compared with just 3 per cent in the USA.

These fisheries are vulnerable to a number of factors.

❑ *Atmospheric and oceanographic conditions.* The atmospheric phenomenon called *el Nino*, for example, caused the Peruvian anchoveta catch to fall from 12 million tonnes in 1970, when it was the largest single fishery in the world, to less than 2 million tonnes in 1973. As the anchoveta catch brought in one-third of Peru's export earnings, the economic consequences were catastrophic. This kind of disruption may become more common with the atmospheric pertur-bations associated with global warming (see p 56).
❑ *Overfishing.* This occurs when too many young fish are taken (growth overfishing), which is inefficient because it fails to exploit their growth potential, or when too many adult fish are taken (recruitment overfishing), which does not allow the species to replenish. There is often a cascade effect with overfishing. As one species or a particular

area is overfished attention shifts to other species or other areas, resulting in further overfishing. Every species has a maximum sustainable yield (see p 85). There are a number of management problems involved in measuring and staying within that limit, which are typical of many natural resource management issues. The life cycle of the North Sea herring, for example, takes it through waters controlled by the UK, Norway, France, the Netherlands, Germany, and Denmark. In this case the European Union's Common Fisheries Policy regulates activities, but there are many instances where there is no such regulatory framework. In addition, single-species management is often unrealistic. A given species will feed on, compete with, or be a food source for various other species. Multi-species management would therefore be greatly preferable, but the necessary knowledge of marine ecology is often lacking.

❑ *Pollution.* Coastal areas and estuaries are especially subject to pollution. Marine life may be directly vulnerable, or may concentrate pollutants in body tissue and thereby become unsafe for consumption, a phenomenon termed biological magnification. The shallow waters of continental shelves are, in general, much more biologically productive than deep ocean waters, and provide much of the global marine harvest. Unfortunately, these waters are also subject to the main flows of pollution.

Mineral mining

Globally, about 27 per cent of crude oil and 20 per cent of natural gas comes from offshore production. Most offshore production is controlled by Saudi Arabia and the UK. Current production from the North Sea is about 90 million tonnes of gas and 160 million tonnes of oil per year. There are also currently some half-dozen aggregate mining sites in the North Sea.

Energy

There are two categories of energy extracted from the ocean:

❑ non-renewable, in the form of offshore oil and gas (see previous section); and
❑ renewable, in the form of waves and tides.

It has been estimated that waves around the UK could supply about 5 GW, about 20 per cent of UK demand. Most of this would come from the north of Scotland, where some of the world's highest wave energies occur. Tidal barrages have an even greater production potential. The proposed Severn barrage, for example, could supply about 7 per cent of UK demand.[31]

Potential conflict for marine resources

There is increasing competition for marine resources, both between countries and between sectoral interests within countries. In the period 1948 to 1976, largely due to decolonialisation, the number of self-determining countries grew from 40 to 150, most of which were concerned to establish claims to resources. The fragmentation of the USSR may lead to new conflicts of this type. Between 1970 and 1977 the ocean area claimed by coastal states grew from 1.9 million to almost 20 million square nautical miles. If every coastal state claimed a 200 nautical mile exclusive economic zone, some 36 per cent of the world's ocean would become sovereign territory.

Water

Table 2: World water

	km^3	Per cent
Fresh water		
Clouds	20,000	0
Continental water	9,000,000	1
Ice	30,000,000	2
Salt water		
Oceans	1,300,000,000	97
Total water	c 1,300,000,000	100

There are more than 1.3 billion cubic kilometres (km^3) of water on the planet, but only 2.7 per cent of this water is fresh (see Table 2). About 78 per cent of the fresh water is frozen, and 97 per cent of the fresh water that is not frozen is groundwater (see Table 3). So just over 0.01 per cent of the water on the planet is surface fresh water.[1, 29]

Groundwater has variable replenishment times. Some groundwater cycles over a period of a few years. Other groundwater cycles over periods of thousands of years, or forms 'fossil' reserves that do not cycle at all. Groundwater that cycles very slowly is effectively a non-renewable resource.

Rain, which is part of an annual cycle, forms a key renewable resource. In the annual hydrological cycle a net 40,000 km^3 of water is transferred to the land as rain (see Table 4). The greater part of this either falls in the sparsely inhabited regions of the planet or runs off as floodwaters.[29]

Table 3: Continental water reserves

	km^3	Per cent
Surface water		
Rivers	2,000	0
Swamps	4,000	0
Lakes	200,000	2
Ground water		
Soil moisture	70,000	1
Subterranean reserves	8,400,000	97
Total continental water	c 8,700,000	100

Table 4: Hydrological cycle

	Evaporation (km^3)	Rainfall (km^3)
Ocean	430,000	390,000
Land	70,000	110,000
Totals	500,000	500,000

Water usage

Fresh water resources fall into three categories:

❑ hydrological cycle;
❑ reserves; and
❑ non-conventional, which includes water obtained by desalination and by treating waste water.

Water requirements also fall into three categories:

❑ domestic;
❑ agriculture; and
❑ industry.

Globally, agriculture is a major consumer of water. It takes some 1,500 kg of water to produce 1 kg of wheat grain, and some 10,000 kg of water to produce 1 kg of rice grain.

As countries industrialise, their main water usage shifts from agriculture to industry. For example, in India less than 10 per cent of water use is domestic and less than 10 per cent is industrial, while agriculture takes over 80 per cent of the total. In Germany less than 20 per cent of the water demand is for agricultural irrigation, while more than 20 per cent is for domestic use and over 60 per cent is for industrial use.

Potential water shortages

Problems arise when supply, represented by stable flows of water, is exceeded by demand. The total available fresh water in regularised or stable flow is 9,000 km^3 per annum. World water use now amounts to 4,700 km^3 per annum (1985 figures). A human population increase to 10 billion by 2100 (see p 74) would increase demand to some 12,000 km^3 per annum. This indicates that demand will outstrip current supply at some point between 2000 and 2050. By 2000 it is likely that some 50 countries will be affected by critical water shortages, with perhaps a further 20 in difficulty, another 20 with 'artificial' shortages (ie without the ability or resources to tap existing water potential), and 10 or more where the shortage problem will be caused directly by contamination of existing water supplies.

Potential conflict for water resources

Watercourses are often used to demarcate borders. This makes political conflict over access to water more likely. Some 148 river basins are shared by 2 countries, and some 52 river basins are shared by 3 or more countries, some by as many as 10 countries. About 20 major rivers form borders. Some 34 countries 'export' water, a further 32 'import' water, and another 16 both import and export water to other countries.

Conflicts to control access to water are likely to become more common. In some cases these conflicts could be resolved through negotiation. In some parts of the world, such as the Middle East, such conflicts might lead to war.[32]

The atmosphere

The current atmosphere of the planet is, to a significant extent, a result of biological activity.

The pre-Cambrian atmosphere was 1.9 per cent nitrogen, 0 per cent oxygen, and 98 per cent carbon dioxide. The first organisms that generated free oxygen were probably the blue-green algae, which can photosynthesise, and which appeared almost 3 billion years ago. Such biological activity has been in large part responsible for moving the atmosphere to its current balance, which is 79 per cent nitrogen, 21 per cent oxygen, and 0.03 per cent carbon dioxide, with traces of other gases.

Atmospheric pollution and global warming

The greenhouse effect may be explained as follows. Some 30 per cent of incoming solar energy is reflected from clouds, particles, and the Earth's surface, while 70 per cent is absorbed at the surface of the planet. The energy absorbed by the Earth is then re-emitted at a lower infrared wavelength (because the Earth is colder than the sun) that can be more easily absorbed by clouds and greenhouse gases. This means that this energy is retained for a longer period in the atmosphere than it otherwise would be (see Figure 8).

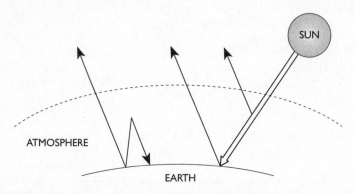

Figure 8: Radiation from the Sun is partly reflected by the Earth and the Sun, but most of it is absorbed by the Earth which then re-emits it in the form of longer-wavelength infra-red radiation. Some of this is absorbed by the greenhouse gases. Thus the Earth's surface and the lower atmosphere is warmer than it would be without the greenhouse effect

The average air temperature at ground level is currently about 15°C. Without a greenhouse effect the temperature would be about 33°C colder, at some –18°C. It is therefore important for many current forms of life that some heat trapping should occur. However, as the concentration of carbon dioxide and other greenhouse gases increases, the heat trapping will increase. If a significant amount of additional heat is retained, the Earth's surface and atmosphere will get warmer, and the climate will change. It is important to distinguish the greenhouse effect per se from the additional or enhanced greenhouse effect which may cause global warming.

The global thermostat

The Earth is kept within certain temperature bounds by its atmosphere. If the Earth had no atmosphere, it would have a similar temperature range to the moon, from –150°C at night to 100°C by day.

At the coldest times during the last two glacial periods, about 20,000 and 160,000 years ago, average surface temperatures were about 3°C colder than today. At the warmest point of the last interglacial, about 150,000 years ago, temperatures were 1 to 2°C warmer than now. This small range is due to the existence of a number of mechanisms that appear to act as a kind of thermostat and maintain the temperature within certain bounds. This thermostat has various geochemical and biochemical components.

The carbonate–silicate cycle, for example, is one important element in this process. Carbon dioxide is washed out of the atmosphere by rain as carbonic acid, which reacts with silicate rocks. Freed ions of calcium and bicarbonate are washed into the ocean where they are taken up by shelled organisms (seashells are mostly calcium carbonate). These organisms die, and their shells form layers on the seabed. Eventually, these layers are forced under the Earth's crusts by movements of the tectonic plates, and the calcium carbonate reacts with silica to form new silicate rocks and release carbon dioxide. The carbon dioxide is then vented back to the atmosphere, especially as volcanos erupt.

This cycle operates as a buffer against short-term change, because the cycle is so long. If the Earth were to get cooler, less water would evaporate, so there would be less rain, and less carbon dioxide would be washed down. If this were to continue for a long time, the cycle would be affected. However, because the cycle is so long, carbon dioxide continues to be vented at an unaffected rate and this will maintain a greenhouse effect. The cycle also contains a negative feedback loop. If the Earth were to get warmer, more water would evaporate, and the additional rain would remove more carbon dioxide, thereby cooling the planet.[23]

There are also some positive feedback loops in these global processes. For example, ice reflects about 90 per cent of incident radiation. As ice caps spread, more heat is reflected, thus cooling the planet and creating more ice. Similarly, if the Earth were to get warmer, more water would evaporate and warm the Earth still further, as water vapour is a greenhouse gas, thereby causing more water to evaporate and so on (unless, as some models suggest would happen, the additional water were to be deposited as snow at the poles). Fortunately, processes such as the carbonate–silicate cycle keep the positive feedback processes on this planet within bounds. The planet Venus appears to have lost all its water, possibly through a similar but unchecked positive feedback process, and now has a surface temperature of 460°C as a result.

It is important to note that the global thermostatic processes are probably robust, but will not necessarily maintain the conditions that are preferred by the human species. The atmosphere has been at very different points of dynamic balance in the past.

There have clearly been transitions between such points of dynamic balance. Some of the phases of these transitions appear to have been relatively rapid. For example, one possible explanation for the Permian–Triassic boundary event some 250 million years ago, at which 95 per cent of all life forms were eliminated, is that the planet experienced an early form of global warming. Essentially, the hypothesis is that a period of glaciation caused a lowering of world sea levels. This exposed extensive coal and black shale deposits, which then eroded and oxidised. This increased the carbon dioxide levels in the atmosphere at the expense of oxygen, to the point where the seas became anoxic. The net global warming effect then exceeded the degree of cooling associated with the glaciation, and the rapid fall in sea levels at the end of Permian times was succeeded by an even more rapid rise in sea levels in the very early Triassic period. The rate of sea level rise at that point may have been extremely fast, possibly as much as several tens of centimetres per year, as large reserves of water were liberated from melting ice.[6]

It is important to note the biological component in these processes. Geochemical and biochemical processes have co-adapted on this planet, and have found a number of points of dynamic equilibrium. For example, when the first photosynthetic bacteria evolved, they probably absorbed atmospheric carbon (which would cool the planet) but released methane (which would warm the planet). As their growth rate depended on temperature, the system would have found an equilibrium. This equilibrium would not have provided an atmosphere that humans could breathe. Subsequent perturbations moved the system to new points of balance, most recently to the one that makes our life possible.

Human impact on the greenhouse effect

There are a number of gases that human activity adds to the atmosphere. A small group of these have the capacity to trap heat with varying degrees of efficiency. However, the relative significance of these gases depends on the amount currently present in the atmosphere as well as their efficiency at trapping heat. Methane, for example, traps heat more efficiently than carbon dioxide but is not as significant as a greenhouse gas as there is currently less of it in the atmosphere. In order of net significance, the gases are as follows:

1. *Carbon dioxide.* This currently accounts for some 55 per cent of the additional greenhouse effect. There are four main global carbon reservoirs: rocks and fossilised carbon deposits, the ocean, the land surface, and the atmosphere (see Table 5).

Table 5: Global carbon reservoirs, in gigatonnes[1]

Reservoir	Carbon (Gt)
Rocks and fossilised carbon deposits	50,000,000
The ocean	10,000,000
Land surface (soil and biomass)	3,600
The atmosphere	700
Total	c 60,000,000

There are flows of carbon between all of these. Carbon flows into the ocean when carbon dioxide is washed down out of the atmosphere. Carbon flows out of the ocean to rocks when calcium carbonate is deposited, and to the atmosphere via respiration by marine organisms. Carbon flows out of rocks to the atmosphere when volcanos outgas or when coal is burnt. Plants absorb carbon by photosynthesis, and release it by respiration and decay. The vegetation cycle usually maintains a balance over twenty-four hours and over a year, with similar amounts of carbon absorbed and released. This is not invariably so, as net increases or decreases in atmospheric carbon tend to accompany periods when global vegetation cover is decreasing or increasing. Carbon is currently being liberated because vegetation is being burnt, cleared, and not allowed to regrow. In addition, global warming may increase respirational losses and accelerate decomposition times, thereby increasing carbon flows.

Human activity mainly affects two of these flows, from rocks to the atmosphere (by burning fossil fuels) and the plant cycle (by burning and clearing trees). This activity currently adds about 5 gigatonnes (billion tonnes, or Gt) of carbon to the atmosphere annually. About 90 per cent of this is due to the combustion of fossilised hydrocarbons, while about 10 per cent is due to the clearing and burning of vegetation.

These activities add an amount equivalent to 0.7 per cent of the total to the atmospheric reservoir each year. Fortunately, only half of this stays in the atmosphere, so that atmospheric carbon dioxide levels are currently rising at some 0.37 per cent per year rather than 0.7 per cent per year, while the rest is probably being absorbed into the ocean and by plants.

As a cumulative result of these and other activities, the carbon dioxide level in the Earth's atmosphere has risen from 275 parts per million (ppm) to 350ppm since the beginning of industrialisation. This is similar to the size of the increase in atmospheric carbon dioxide at the end of the last two glacial periods, when levels went up from 190ppm to 280ppm.

The Intergovernmental Panel on Climate Change (IPCC) have estimated that emissions could reach some 20 Gt per year by 2050 if current trends are continued, but that with a major shift in a number of relevant policies, additional emissions could be reduced to some 2Gt per year by 2050.[33] However, as global consumption of fossil fuel is currently rising at some 1.5 per cent per year (varying from 0.5 per cent in the industrialised nations to 6.0 per cent in the developing nations), the low emission scenario currently seems less likely.[23]

Most of the carbon in the land surface reservoir is in the top metre of soil. Carbon could be liberated from this organic soil matter by microbial biomass if temperatures increase and promote increased rates of decomposition. As a result of the global warming predicted to occur over the next 60 years, an additional 61 Gt of carbon dioxide could be released from the land surface reservoir in this way, thus adding to the greenhouse effect.[34]

Such effects, which may be triggered at certain threshold points as systems transform, are termed 'enhancers' when they would have the effect of accelerating or 'enhancing' an existing trend.

2. *Methane.* This currently accounts for some 15 per cent of the additional greenhouse effect. Methane levels have more than doubled since industrialisation, mainly as a result of rice and cattle cultivation,

biomass burning, and gas leakage from oil and coal recovery.[23] The role of methane may become particularly significant. There is a possibility that methane could become involved in a positive feedback loop as an enhancer. If temperatures rise sufficiently, quantities of methane could be released from the large reserves currently frozen in the Arctic tundra, in an analogous way to that in which some of the soil reserves of carbon would be liberated at a certain stage of global warming. If the tundra should start to defrost, and the reserves of methane were liberated, this would tend to increase any warming effect.

3. *Chlorofluorocarbons (CFCs)*. These are industrial products, used in solvents, refrigerants, and propellants. They were not released in any quantity until the 1950s. Their contribution to global warming is currently disputed. They are very efficient at trapping heat, and, if this effect predominates, they could be accounting for some 24 per cent of the additional greenhouse effect. However, the chlorine in CFCs catalyses the breaking-down of ozone. This has a number of consequences (see p 65), one of which may be a diminished absorption of solar energy and consequent cooling of the atmosphere.

4. *Other greenhouse gases* include water vapour, which plays an important role in atmospheric dynamics, nitrous oxide, which may account for some 6 per cent of the additional greenhouse effect, and ozone.

There are other atmospheric pollutants which may be causing a global cooling effect (see p 67), and thus masking some of the global warming effect. If this is correct, then rates of global warming would increase if emissions of these pollutants were reduced.

Global warming projections

Current estimates are that the planet may have warmed by 0.5°C over the last century and that, without prompt and effective intervention, the planet could warm by as much as a further 1.5°C to 5.5°C over the next 150 years. If carbon dioxide levels double by 2050 or 2100, as they are currently projected to do, then the average temperature might rise by some 3°C. It is possible that this average temperature rise will be distributed unevenly. Some models, for example, indicate that the rise will be greater at the poles than at the equator.

Changes in sea level are likely to accompany any significant increase in the average temperature, for various reasons:

❑ The thermal expansion of the ocean;
❑ Some melting of polar ice and continental glaciers;
❑ The worst-case scenario – the possible disintegration of the West Antarctic ice sheet.

Current projections suggest that a temperature rise of 3°C could cause a rise in sea level by 0.3 to 1.2 metres, most of which would be due to the thermal expansion of the ocean. The most recent estimates, fortunately, indicate that the sea level is unlikely to rise by more than 0.3 metres by the year 2100, due partly to the cooling effect of other atmospheric pollutants referred to on pages 59 and 66. Any further warming, however, would accelerate this rise in the sea level.

If the sea level were eventually to rise by 1 metre, this would affect (by inundation, salination and so on) some 5 million km^2 of land, which is about 3 per cent of the land surface of the Earth. This would affect urban and agricultural land disproportionately, for several reasons.

❑ Many of the world's great cities have grown around harbours and sea-ports, and lie near sea level.
❑ Fertile land tends to be low-lying, in valley bottoms and river deltas. The 3 per cent of the Earth's land surface that would be vulnerable to a 1 metre rise currently includes about a third of the world's most productive cropland. As vegetation belts (see below) may shift with increasing temperature, agricultural production may have to be relocated anyway, but this would probably have to be to areas with less suitable soils.

Most models suggest that the global weather and tidal systems will be significantly affected. Current indications are that the total global rainfall might increase by about 10 per cent, but also that global warming will probably change the distribution of the rainfall around the globe, so that some regions of the globe will actually become drier. The weather will probably become more variable, with increased occurences of extereme weather, such as cyclones, with consequent storm damage and flooding.

Western Europe has a relatively mild climate, which is largely due to a circulatory system called the Atlantic conveyor. Warm water moves north from the equator, giving off heat as it does so. It is cold by the time it has reached high latitudes. Cold water is more dense, so the water sinks down at the high latitudes and returns to the equator in the depths of the ocean. As the water sinks, it pulls part of the warm surface water in the Gulf Stream north, so establishing a giant convection current. Fresh water, which is deposited into the ocean as rain or snow, makes seawater less dense. A large influx of fresh water from increased rainfall would therefore act as a 'brake' on the conveyor. One study indicates that a doubling of carbon dioxide levels, which is currently projected to occur by about 2065, would significantly weaken the conveyor – and that a quadrupling of carbon dioxide levels would shut it down.[35] This would probably have a marked effect on Western Europe's weather. It could even mean that the UK would eventually become colder as a result of global warming.

The effects on the poles are less clear. Some models suggest that some of the polar ice will melt. Others suggest that part of the increased rainfall will come down as snow on the poles and add to the thickness of the ice. Recent work on Greenland suggests that both effects are happening, with the ice thickening on the west coast of Greenland (which is exposed to the snowstorms carried on the prevailing westerly winds) and melting on the relatively sheltered east coast of Greenland.[36] Which effect will predominate remains to be seen.

Any significant changes in weather and rainfall would affect the distribution and supply of fresh water (see p 54), which would have implications for industry and agriculture. One projection suggests that the American mid-West will become drier, which would reduce crop yields. This would have global significance, as North America currently produces some 87 per cent of the world's exported grain.[37]

Western Russia and Ukraine, on the other hand, may be able to increase production of grain, as some projections indicate that they may get a more favourable climate as a result of global warming. There may be a number of such 'compensating' effects, but it is important not to disregard the political import of such changes. If the Ukraine replaced the American mid-West as a major grain-exporting region, for example, the political repercussions could themselves lead to a period of instability in food supply.

Over the last million years the Earth has gone through a regular cycle. This cycle, which is driven by slight irregularities in the Earth's orbit, results in glacial periods of about 100,000 years interspersed with warmer interglacials of about 10,000 to 20,000 years. The current interglacial is about 12,000 years old. Since the peak low temperature of the last glacial period the temperature has risen by about 3.0°C, but has taken about 20,000 years to do so. The projected rise due to the greenhouse effect will occur, therefore, about 100 times faster.[23] This rapid rate of warming may prove to be a particularly important factor, as it is likely to affect species extinction rates.

The distribution of vegetation over the planet is dynamic. Fossil records indicate that the vegetation distribution shifts some 200 km towards the poles with each 1°C rise in temperature, and that the forest migration rates at the end of the last glacial period were some 20 to 100 km per century. Conditions would gradually become less favourable at the southern bound, so that fewer seedlings would survive, and more suitable at the northern end of the range, so that more seedlings would survive. Thus tree belts would slowly move northwards.

The effects of slow global warming, therefore, can be absorbed by the forests. If future global warming proves to be some 100 times faster than the rate of warming at the end of the last glacial period, however, then many tree species would be unable to 'migrate' at the necessary speed. This would affect any dependent species (see p 38).

It must again be emphasised that all such projections, estimates and scenarios are still highly uncertain. The actual outcome could be much less severe than suggested above if, for example, the global processes that help to maintain the current balance prove to be sufficiently robust. However, it is also possible that the actual outcome could be more severe than suggested earlier.[38] The most serious scenario is that positive feedback effects might start to predominate, creating further global warming and, eventually, leading to a situation beyond human control or ability to take remedial measures.

Countering global warming

Various methods have been proposed for reducing the rate of global warming. Increases in energy efficiency and reductions in rates of deforestation are clearly essential. It has also been suggested, however, that more could be done to re-absorb some of the excess atmospheric carbon. Reafforestation is one way in which this could be achieved. A mature forest gives off about as much carbon as it absorbs, but a new plantation acts as a net carbon sink because the trees take up carbon as they grow (although this carbon would be re-released if the trees were then cut for fuel). Given the other beneficial environmental effects associated with sympathetic tree-planting, and the scope for extensive reafforestation in many parts of the world, this is clearly an attractive option.

More recently, it has been suggested that marine algae (phytoplankton) could be used to re-absorb atmospheric carbon. There are large regions of ocean – perhaps 20 per cent of the total – that are rich in essential nutrients but sparsely populated with phytoplankton. Lack of iron, which is important for cell growth, appears to be the limiting factor. 'Fertilisation' of these areas of the ocean with iron sulphate could therefore encourage a spurt of algal growth. Phytoplankton take up carbon when they photosynthesise, so a significant increase in their number would absorb worthwhile amounts of atmospheric carbon.

Early experimental results have indicated that a dramatic increase in the number of phytoplankton does occur when iron is added to the ocean, but also that this effect appears to be limited by two factors. One is that much of the iron gradually sinks below the surface waters where the algae live. The other is that the increase in the number of phytoplankton is rapidly followed by an equally dramatic increase in the number of zooplankton, which eat the phytoplankton, thereby re-releasing the carbon via respiration and organic breakdown.[39] More recent work, however, suggests that it may be possible to overcome these two limiting factors by adding the iron in repeated doses although many more years of research will be needed before these processes are sufficiently well understood to become predictable.[40]

This 'technofix' might, if applied, help to counter the problem of global warming. There are, however, a number of important issues involved in the choice of such a solution. It is possible, for example, that a concentration of effort on a search for solutions to symptoms – as opposed to causes – could have the effect of directing attention and resources away from avenues that might lead to more fundamental, benign or permanent solutions. The use of iron to encourage algal blooms and the rapid sequestration of atmospheric carbon could lead, for example, to a general perception that the global warming problem has been solved, and that we can therefore continue to tolerate inefficient usage of fossil fuel. This would remove some of the incentive to invest in energy efficiency, encourage the continued depletion of remaining fossil fuel reserves, and thereby prolong the exposure of sensitive ecosystems to the acid deposition also associated with fossil fuel use.

A general problem with technofix approaches is that the 'solutions' generated sometimes turn out to have unintended or undesirable side-effects, which then require further costly action. It might, in the long run, be more cost-effective to address the problems at source.

Notwithstanding these reservations, such applied research is likely to extend our knowledge of the behaviour of the environmental systems in question, and there is always the possibility that the research will actually lead to a solution that genuinely represents the best available option. It would be unwise, however, to assume that such technofixes will always be available.

Ozone depletion

Ozone is distributed unevenly through the atmosphere. About 90 per cent is in the stratosphere, and most of the rest is in the troposphere. These are, respectively, the second-lowest and the lowest layers of the atmosphere. The highest ozone concentrations are at altitudes of 20 to 35 km, and this is called the ozone layer (see Figure 9).

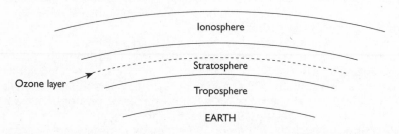

Figure 9: The atmosphere consists of a number of layers with different characteristics. The 'ozone layer' is part of the stratosphere

Ozone is constantly created and destroyed in the stratosphere. The process is very complex, involving between 100 and 200 interactions. Essentially, what happens is that oxygen molecules (which consist of two atoms of oxygen) absorb ultraviolet (UV) radiation. This gives them enough energy to split into two separate oxygen atoms. Both of these can then bond with another oxygen molecule to form ozone (which consists of three atoms of oxygen). This process absorbs, and so filters out, most of the incoming short wavelength UV radiation. Later on the new ozone molecule, by absorbing a slightly longer wavelength UV, splits up again. Eventually, one of the free oxygen atoms meets an ozone molecule and the cycle ends, with the four oxygen atoms recombining into two oxygen molecules.

This highly complex process, under normal conditions, achieves a dynamic steady state, in which the rate of ozone creation is balanced by the rate of ozone destruction.

There are a number of substances that act as catalysts in this series of reactions. That means that they enable some part of the process to happen much faster. Chlorine is especially effective. Every free atom of chlorine in the stratosphere destroys about 100,000 molecules of ozone before being removed or inactivated.

Catalysts are an essential part of the process that balances the ozone budget. However, current human activity is generating additional flows of some of these catalysts, and some 80 per cent of the current chlorine loading of the atmosphere was added by human activity.[41]

Chlorofluorocarbons (CFCs) contain chlorine. CFCs are very stable, and do not break up until they reach an altitude of 25 to 40 km and are subject to high levels of UV radiation. They then liberate chlorine.

This chlorine, given the right kind of atmospheric conditions, will then destroy ozone. These atmospheric conditions, essentially, are strong circular winds to hold a column of ozone steady, with low enough temperatures to form ice crystals to provide a surface on which chemical reactions can occur, with enough UV to break up the CFC's. The reason that ozone depletions have been observed originating over Antarctica rather than over the Arctic is that the right weather conditions, the strong circular polar vortex winds in particular, occur regularly over the Antarctic pole and less regularly over the Arctic pole.

Ozone-poor air then flows out when the onset of summer ends the polar vortex. Over the period 1969 to 1986, ozone levels in the winter months decreased by 2.3 per cent to 6.2 per cent between latitudes 30° North and 60° North. The largest losses were in the higher latitude bands. Depletions of 5 per cent have been measured north of 40° North (which includes Europe, Russia, Canada, and the US), and depletions of 7 per cent have been recorded north of 53° North (which includes all parts of the UK north of Wales). Even more substantial ozone depletions were recorded in January 1992, when short-lived peak depletions of 45 per cent were measured over the UK and Scandinavia. The most dramatic depletions are, of course, over Antartica, where they originate. Ozone depletions of 95 per cent have been recorded over parts of Antarctica.

Effects of ozone loss

As ozone is depleted, less of certain wavelengths of UV will be absorbed, and more will therefore reach the surface of the planet. There is an especially large effect with the wavelength known as UV-B, the amount of which reaching the surface increases by 2 per cent with every 1 per cent reduction in ozone. This could have a number of adverse consequences.

❑ The Environmental Protection Agency in the US calculates that every 1 per cent reduction in ozone, by allowing additional exposure to UV-B, will cause a 3 per cent rise in cases of non-melanoma skin cancers, which would mean some 12,000 to 15,000 additional cases in the US each year.

❑ Indications so far are that some two-thirds of mid-latitude agricultural crops are sensitive to increased UV-B, and typically respond with reduced growth and yield.
❑ The wavelength UV-B penetrates 5 to 20m into water, depending on the turbidity. The greater part of the marine biota live near the surface. Indications are that increased UV-B damages a wide range of marine organisms, some of which form the base of marine food chains. Impacts here may therefore have a cascade of consequences up through food chains and rapidly affect fisheries (see p 50).

There are also climatic implications of ozone loss. The ozone cycle absorbs solar radiation. Some of this energy is eventually given off as heat into the stratosphere. This produces a temperature inversion at the top of the troposphere, which is what maintains the stratosphere. Ozone loss, by cooling the stratosphere, may change the temperature structure of the atmosphere. This may, in turn, increase the occurrence of the weather conditions in which ozone is destroyed, thus establishing a dangerous positive feedback loop.

The Montreal Protocol of 1987, with subsequent revisions, commits signatory nations to phasing out CFCs and certain other chlorine compounds by 2000 or 2005. However, as CFCs are so long-lived, their effect will persist for many decades to come. It is also unfortunate that some of the compounds that have been developed to replace the CFCs, such as the hydrochlorofluorocarbons (HCFCs), also contain chlorine, albeit in a form that has a shorter atmospheric lifetime, and so will make some further contribution to total atmospheric chlorine loading. Current indications are, fortunately, that the use of HCFCs as a substitute for CFCs will only be permitted as a temporary measure.

Acid deposition

The greater part of acid deposition is in the form of oxides of sulphur and nitrogen. The main sources of additional sulphur dioxide emissions are fossil fuel-burning power stations, while the main sources of additional emissions of oxides of nitrogen are motor vehicle engines. These emissions are vented into the atmosphere, then return to Earth in wet or dry form. Much returns in rain as sulphuric and nitric acid. This acid deposition damages certain vulnerable ecosystems. Direct economic costs include acid damage to buildings, machinery, freshwater fisheries, agricultural land, and forests (see p 68).

Sulphur dioxide may also have a global cooling effect. It is possible that it helps to reflect some solar radiation. It is therefore possible that this effect is currently mitigating the global warming effect caused by other pollutants (see p 59). This means that any restriction of emissions of sulphur dioxide in order to reduce acid deposition might reveal that the underlying global warming trend is stronger than is currently apparent.[42]

This is a good illustration of a general point. It is argued here in a number of places, including at the beginning of this section, that large or fast change is inherently more risky than small or slow change, although small or slow change is not necessarily risk-free. This is generally true for ecological and other complex systems, which might be able to adapt to change provided that the rate and extent of the change is not too great, but which are more likely to be catastrophically affected by large or fast change. For example, tree belts could migrate with slow global warming, but would probably die back with rapid global warming (see p 71).

It is perhaps less obvious that this principle can actually be true of remedial measures too. It may be better, in some circumstances, to reform cautiously. It is possible that even the process of reform could induce adverse consequences. This is more likely to happen in those instances where a number of balances have been disturbed simultaneously, and other biological or chemical systems may have moved to some new point of balance that itself is now being maintained by some human input. This factor should not, of course, be elevated into a rationale for comprehensive inaction. A rather more satisfactory solution, of course, would be to place all such reforms in the context of an integrated package of remedial measures aimed at a number of the dimensions of an overall problem.

Land use: agriculture and soil erosion

About one-third of the Earth's land surface is too arid or too cold to be used for any kind of cultivation. The area currently in production could be increased, possibly even doubled, but the cost of doing so would rise rapidly as more marginal areas were brought into use (see Table 6).

The land currently in arable production is of variable quality. The developed countries produce some 50 per cent of world agricultural output with only 25 per cent of the world population and 7 per cent of the world agricultural labour force. This is for a number of reasons. These nations possess nearly 50 per cent of the world's high quality arable land (Europe has a higher percentage of land in arable production than any other part of the world), and have more advanced technology. Another significant factor is that these nations can afford high levels of energy and other resource inputs. Between 1950 and 1980 fertiliser use in the developing nations increased from 1.4 kg nutrients per hectare (n/h) to 49 kg n/h. In the developed nations, however, over the same period, fertiliser use increased from 22.3 kg n/h to 116 kg n/h.

Some 150 years ago a farm in a developed nation would demonstrate a ratio of purchased inputs to gross value of outputs of the order of 5 per cent. The figure today would be nearer 50 per cent. This is because modern rice production methods, for example, require some 375 times as much commercial energy input per hectare as traditional methods. Of course, yields are higher, but the modern production methods still require about 80 times as much commercial energy input per hectare and per kilo of rice

Table 6: World land use by region, FAO 1985 figures, in million hectares, percentages in parentheses

Region	Area	Arable	Pasture	Forest	Other
Europe	472	139 (29.4)	84 (17.8)	155 (32.8)	94 (20.2)
N America	1838	236 (12.8)	272 (14.8)	591 (32.2)	738 (40.2)
S America	2054	180 (8.8)	553 (27.0)	984 (47.9)	337 (16.3)
Oceania	842	50 (5.9)	453 (53.8)	159 (18.9)	179 (21.4)
Former USSR	2227	232 (10.4)	374 (16.8)	935 (42.0)	687 (30.8)
Africa	2964	184 (6.2)	788 (26.6)	697 (23.5)	1295 (43.7)
Asia	2678	454 (17.0)	644 (24.0)	562 (21.0)	1018 (38.0)
World Total	13075	1475 (11.3)	3168 (24.2)	4083 (31.2)	4345 (33.2)

produced as do traditional methods. Such factors mean that in the US, for example, the energy cost of food in 1970 was 5 times higher than in 1900.

The biological productivity of the planet is powered by solar energy. Plants utilise this energy directly. Animals cannot, so must eat plants to capture this energy. Some animals capture energy by eating animals that eat plants. Every time the energy is converted in this way there are conversion losses. A typical conversion loss in this process would be about 90 per cent. This is why herbivores can exist in large numbers (because they can obtain energy directly from plants), and why carnivores can only ever exist in small numbers (see p 81). The current levels of human population, which are more characteristic of herbivores than carnivores, are made possible by the fact that humans are omnivores, and base much of their diet on plants. This means that the current level of human population has been made possible by the global increase in agricultural productivity.

The current levels of agricultural production have in turn been made possible by high levels of energy and other inputs. Most of this energy has been derived from fossil fuels. Many of the other inputs (either directly, in the case of some petrochemicals, or indirectly, in manufacture) are also derived from fossil fuels.

One important implication is that the sustainability of the high energy input–output mode of agricultural production will depend, at least in part, on the sustainability of energy production (see p 91).

Agriculture and other forms of human land-use can induce land degradation. This can take different forms, including soil loss through acidification, alkalisation, salination, denaturing, wind or water erosion, and desertification. Some forms of degradation, such as gullying, are fairly obvious; others, such as chemical changes in the soil, may not be immediately apparent.

❑ *Acidification.* Most crops need soils in the range pH7 (neutral) to pH5 (slightly acid). Many soils are naturally acidic. However, further acidification results from emissions of sulphur dioxide and nitric oxides (see p 67). Excessive soil acidity slows growth. There is a partly compensating effect, however, with emissions that contain sulphur, nitrogen, or ammonium, as these chemicals can promote growth. Overall, acid deposition probably reduces yields. In parts of eastern Canada, for example, acid rain is estimated to have cost some 10 per cent reduction in crop yields.

❑ *Salination.* Salination of soils, especially in semi-arid areas, is often linked with irrigation. Irrigation can raise the water table until ground salts are brought up to the root zone and stunt crops. It has been suggested that approximately 50 per cent of the world's irrigated crop land is affected in some degree by salination.

❑ *Denaturing.* Denaturing, or loss of soil structure, is associated with loss of organic matter from the soil. It can be caused by cultivation of excessively wet soils. It usually leads to structural instability, slumping, surface compaction, decreased infiltration rates, increased run-off, reduced cation exchange capacity, which tends to increase the loss of nutrients through increased leaching, and raised rates of erosion.

❑ *Erosion.* Soil erosion is part of a natural cycle, in which soil is created, cycled, and lost. Problems arise when the soil erosion rate is pushed above the soil formation rate. This can happen, for example, when vegetation cover is removed or steep slopes are farmed. The soil formation rate in the UK and the US averages between 0.02 mm and 0.1 mm per year, which gives 0.3 to 1.3 tonnes per hectare per year (t/h/y). The US Department of Agriculture has estimated that the maximum erosion rate that any soil type can withstand, while still permitting high levels of productivity at economically viable levels of inputs, is around 11.2 t/h/y. Some 27 per cent of all cropland in the US is currently eroding faster than this, and some 10 per cent is eroding at double that rate or more. Apart from the consequent long-term loss of productivity, soil erosion also causes downstream silting-up of water courses, which in turn causes flooding and shortens the life of reservoir and hydro-electric schemes.

❑ *Desertification.* Desertification is the final stage of a process of land degradation in the semi-arid parts of the world. It is usually achieved by a combination of factors. These may include decreased rainfall or increased variability of rainfall (both of which may become more likely with global warming; see p 61). Appropriate land-use patterns could assist in maintaining the vegetation cover necessary to prevent soil loss. Unfortunately, more typical land-use patterns include overgrazing and consequent loss of vegetation cover, followed by soil loss and gully erosion, salination, and alkalisation.

The UN Food and Agriculture Organisation (FAO) estimated in the early 1980's that, as a result of all of the factors described in this section, some 25 per cent of the world's arable land was being degraded. This estimate was subsequently revised upwards, to some 33 per cent. Related estimates have suggested that some 7 per cent of the world's cropland was lost to soil erosion between 1980 and 1990, and some estimates of the current global topsoil loss rate are as high as 24 billion tonnes per annum. Others, however, have argued that these estimates are too high and, as a general rule, all such estimates must be regarded with caution. It should be noted, however, that the loss rate in any one year will be a fraction of the soils rendered vulnerable to erosion, which suggests that considerable effort would be required to stem this rate of loss, and that these losses in agricultural productivity are partly masked by the increased levels of inputs referred to earlier, which means that any reduction in agricultural inputs may now have more serious implications for food supply than would have been the case even as little as fifty years ago.

Deforestation

Human activity has been gradually deforesting the planet over thousands of years. The temperate regions of Europe and North America were cleared first. The greater part of Britain, for example, was forested until some 7,500 years ago. By the beginning of this century, UK tree cover had fallen to 5 per cent.[43] Today, the country has about 10 per cent tree cover, but retains mere fragments of primary forest.[44] Since the early part of this century most deforestation has occurred in the tropics. The tropical rainforests have been reduced by about 40 per cent in the period since 1945. Between 11 million and 15 million hectares of tropical forests are currently cleared each year, with a further 10 million hectares degraded.

This process is driven partly by demand for timber from the developed nations. Japan is the largest timber importer in the world. With 2 per cent of the global population, Japan consumes 40 per cent of the world's traded mahogany. Britain too is a very significant importer of timber.

The process of deforestation has a number of ecological implications. There are likely to be consequences for the atmosphere (see p 59), because forests store 20 to 100 times more carbon per unit area than cropland, and because the burning of the timber is currently adding some 0.5 billion tonnes of carbon dioxide to the atmosphere.[45] There are also likely consequences for the rate at which genetic diversity is lost (see p 72), because forests typically support many other forms of life. Rainforests are particularly significant in this regard because they probably contain at least 50 per cent of all non-marine life forms, although they only cover 6 per cent of the planet's surface.

Global warming could affect the extent of forest cover in various ways. For example, one of the effects of certain levels of global warming may be to increase the rate of plant growth due to raised levels of atmospheric carbon. This could assist in reafforestation. It is also possible, however, that rapid rates of global warming would cause significant die-back and loss of tree cover, because tree belts would be unable to migrate at the necessary speed (see p 61). In time, it should become apparent which of these effects will dominate.

Biodiversity and extinctions

Gradualist theories suggest that all the observed diversity of life can be accounted for by fairly uniform rates of species evolution.[21] The (controversial) theory of punctuated equilibria indicates that species evolution rates vary significantly over time.[46] There is a more general consensus, however, that extinction rates are irregular. Research into system dynamics indicates that complex adaptive ecological systems can demonstrate large intermittent fluctuations, interspersing periods of relative stasis with periods of rapid change.[47] This means that some irregularities are part of the internal dynamics of the system. Other irregularities, or ecological system state transitions (see p 31 and p 38), may be caused by significant external or extraterrestrial perturbations. One possible explanation for the mass extinction at the end of the Cretaceous period, for example, is that the Earth was hit by an asteroid or planetoid.

The rate of species extinction may therefore result from two factors – a low rate of gradual 'background' loss with intermittent mass extinctions (defined as the loss of more than 5 per cent of the total number of species extant in any single event) reflecting more serious system state transitions.

The latter would have the effect of superimposing a number of peaks on an otherwise fairly regular rate. It has been estimated that over the last 600 million years, since multicellular life forms evolved, background extinction has cumulatively accounted for some 40 per cent of all extinctions, while major transition events have accounted for some 60 per cent of all extinctions.[48]

The number of species extant is not known. About 1.7 million have been classified. However, there is a general consensus that many more remain unclassified, and estimates of the total range from 2 million to 80 million. This, obviously, makes it difficult to estimate the present rate of loss, because it is likely that currently unknown species are also becoming extinct. Estimates of the rate of loss consequently vary, and this is another fairly contentious area of research. Most currently agree, however, that the largest contribution to the overall loss probably results from the progressive loss of the rainforests, and a recent (and respectably conservative) estimate is that the loss which results from rainforest clearing alone is between 4,000 and 6,000 species per year.[49]

Of course, species extinction is a essential part of the process of evolution. However, it is important to note that the current extinction rate – on the basis of estimates such as those above – is about 10,000 times faster than the usual background rate. This means that the current rate probably resembles those at major extinction boundaries rather than the more usual background rates.

This rate of loss is different from previous mass extinctions, in that it results from voluntary activity by one species. The same pattern of voluntary activity is also depleting the biological 'infrastructure' of topsoil and vegetation, which may delay any recovery.

The loss of diversity within species can be a more immediate problem for humans than the loss of species. The greater part of the human diet consists of some eight crops. There are only 20 major crops in production in Europe and North America, and only six major domesticated species of animal. It would be perfectly possible to change and widen this diet, over time, but at the moment we have a relatively specialised diet, at least in comparison to the range of foods that could be eaten. There are, of course, a number of historical, cultural, economic and practical reasons why we have specialised in this way. The important point is that the continued viability of this specialised diet has depended, at various points in the past, on a wider genetic reservoir from which fresh genetic material could be drawn on occasions when particular domesticated strains of plant and animal species became vulnerable to variant viruses, insects, or fungi. The loss of genetic diversity within species represents the loss of that genetic reservoir. The significance of that loss might not be apparent until some particularly virulent variant of a major pest appears and threatens to destroy a crop. Genetic engineering might, in time, be able to substitute for some part of these losses. At present, however, no technological solution could provide an even partial substitute for diminished genetic diversity.

It is also important to note that some technological solutions create further problems. Some agrochemicals used to control pests, for example, have the long-term effect of encouraging the most chemical-resistant strains. They do this by, in effect, changing the evolutionary parameters to favour such resistant strains. Genetic engineering appears to offer some attractive solutions, but it is possible that the release of transgenic organisms into the environment may create problems as or more serious than those solved. The substitution of technology for ecological balance which, unfortunately, is being forced and accelerated by commercial interests, may therefore not be a good long-term strategy (see p 100).

The loss of a species changes the ecology of an area. It is arguable, however, that species do not all play equally important roles in maintaining particular ecological balances. The loss of one species might have only a limited impact, while the loss of another 'keystone' species might have an extensive impact and cause a cascade of further losses.

There can also be cumulative and delayed effects. A gradual reduction in genetic diversity, for example, might not matter as much during periods of relative system stability, but could reduce the overall ability of the ecological system concerned to respond flexibly to stress (see p 31). The real significance of the loss of the genetic diversity may not become clear, therefore, until after the onset of some crisis.

The loss of natural checks and balances could therefore incur a number of adverse consequences, ranging from relatively minor to very extensive effects. These might affect the human species in a number of ways. Some losses might have no perceptible impact on the human species. Some might incur an indefinite commitment to manage an ecology in order to prevent a more serious collapse which could threaten an industry, or cause widespread social and economic damage. Some might even threaten the survival of the human species.

Population

For most of human history there have been only a few million people alive at any one time. Levels have probably fluctuated, but reached perhaps 300 million some 2,000 years ago and about 600 million 500 years ago. Numbers have increased steadily since the late 17th century. In 1987 the world's population exceeded 5 billion.

The rate of increase now appears to be exponential. The population took 1,500 years from the year 0 to 1500 AD to double to 600 million, just 150 years from 1750 to 1900 to double to 1.7 billion, and only 30 years from 1950 to 1980 to nearly double to 4.8 billion. The world's population is currently 5.2 billion (as at 1990), and increasing at 2 per cent each year. An annual growth rate of 2 per cent means a current recruitment rate of about 100 million people each year, and a doubling period of 35 years.

Indefinite population increase is not possible. At current rates of growth, there would be standing room only by about 2500. Global resource flow potential and pollution absorption capacity would probably be exceeded considerably before that point.

However, further population growth currently appears inevitable. This is partly due to the age structure of the existing population (a relatively high percentage of the population in many developing nations is young), and partly due to inadequate political will and a lack of the necessary mechanisms, such as education programmes and generally available contraception. One set of UN estimates indicates that if the rate of increase was reduced to replacement levels by 2010, the population would stabilise at 7.7 billion by 2060, but if this reduction were not achieved until 2035, the population would not stabilise until it reached 10.2 billion in 2095. If it took until 2065, the population would reach 14.2 billion by 2100. More recent estimates from the International Institute of Applied Systems Analysis (IIASA), with more sophisticated modelling of the underlying factors of fertility, mortality and migration rates, indicate that the world's population will increase by 50 to 100 per cent by 2030. The IIASA mid-range scenario, that currently believed to be the most likely outcome, is that the world's population will be 12.5 billion by 2100.[50]

These population increases will require a concomitant increase in economic activity, which in turn is likely to lead to an increase in the demand for resources and increased generation of pollution. The human population will eventually stabilise at some stage, at a considerably higher level than today. This stabilisation may actually generate additional problems, as the population will then gradually become more elderly, so that the economically active members of the population will have to support larger numbers of elderly dependents.

There is a critical distinction between global and regional limits to population growth. There are extreme inequalities in the current global distribution of population density, mineral and energy resources, and biological productivity. This situation is made possible by interactions and exchanges between regions, economies, and societies, which transfer resources between parts of the planet (see p 90).

There are also very significant geographical variations in the rate and character of population change. These are due to a highly complex interaction of economic, social, cultural, and political factors with mortality and fertility rates.[51]

One important factor is the net direction of inter-generational wealth flows within the household. There are pro-natalist incentives where children are cheap (where there is, for example, no requirement to educate children) and their return is high (where children are able to earn wages, or supply labour for agriculture). There are anti-natalist incentives where children are expensive (where there are, for example, extended educational requirements) and their return is low (where children are not allowed to work). Another important factor is the relationship between sexual differentiation of labour and the sexual differentiation of power. In parts of Africa, for example, the traditional sexual differentiation of labour ensures that the immediate results of increased population and consequent pressure on environmental resources bear more heavily on women. Political and economic decision-making power, however, tends to be concentrated among men. These disparities can lead to a misallocation of resources and an inability to respond to changing demands.

Of the 100 million people added to the population each year, some 80 per cent are in developing nations. By 2030, developing countries will represent some 85 to 87 per cent of the world's population. Many of these countries do not currently have the surplus production capacity or expansion potential to support this increasing demand. For example, at a global level, growth in cereal production has outstripped population growth. Yet, as a result of a complex interaction between a number of factors, including patterns of domestic land ownership, international trade, international debt and migration, food production failed to keep pace with population growth in 70 of the 126 nations in the decade 1970 to 1980.

One of the first demographers, Malthus, believed that populations could increase geometrically, whereas agricultural productivity could only increase arithmetically. In societies where labour was sold as a factor of production, population growth would lead to rising food costs (due to increased demand) and falling wages (as more labour came on to the market). This would lead to widespread malnutrition, and thus to rising mortality. Malthus could not anticipate, however, the way in which technology, coupled with social and economic reorganisation, would transform the productive potential of the economy and so relax some of the constraints imposed by the physical environment. In this way, the population has been able to escape the Malthusian restrictions, at least to date.

Given that nearly one-fifth of the existing world population is malnourished, however, and given the anticipated rate of increase in the population, it is hard to imagine the nations of the world coping particularly well with the task of providing food, clothing, housing, fuel and water for all the additional people, let alone absorbing them all into the world's labour markets. It is also hard to see how such growth can be reconciled with widening aspirations towards highly consumerist and energy- and resource-intensive lifestyles.

5

Ecosystem Economics

This section reviews some of the complex ways in which economic and environmental systems interact. There are a number of ways in which environmental features constitute or enable flows into human economic systems. These inputs are not all equally important, in that we are currently more dependent on some than others, or equally critical, in that depletion of some factors would have relatively confined consequences, while the depletion of others could have wide implications and cause significant additional impacts and losses.

Some inputs are renewable, so could, with good management, be available indefinitely. Others are non-renewable, so could, in principle, be depleted to exhaustion. This might or might not be significant, depending on how extensive the reserves are, how well they are managed, and what steps are taken to reduce the dependence on the resource concerned.

There are also a number of ways in which human economic systems generate flows of various types into environmental systems. Some of these can probably be easily absorbed, while others may exceed in various ways the ability of the environmental system concerned to absorb them without causing some potentially adverse ecological change.

International trade means that such inputs and outputs flow between different political and economic systems and operations, so some of these effects are displaced geographically, or temporally, or both.

Humans depend on the products of economic processes, such as agriculture and industry, that convert environmental inputs into economic outputs. There are various factors that limit the extent of these conversions, some of which may become more important as human demands increase. If this conversion process has to be rather more carefully managed in future, there are a number of ways in which this might be achieved, each of which would have various consequences.

Definitions of capital

The word 'capital' is used throughout this book in different contexts. The meanings are as follows:

The word *capital* is used to refer to human-generated wealth, including both the goods that society produces and the tokens of value, such as money, that society uses to transfer wealth.

The phrase *artificial capital* is used to refer to the goods that society produces, in contrast to natural capital. Less tangible factors of production, such as services and a knowledge base, are also taken to be forms of artificial capital. The terms 'reproducible' or 'man-made' capital have been used by others to describe this kind of capital. These terms have not been employed here. In the case of the term 'reproducible', this was to avoid possible confusion between the idea of reproducible (artificial) capital and renewable (natural) capital. In the case of the term 'man-made', this was to avoid the connotation that the source of capital is human. In practice, the source of most of this capital is natural, but its form has been artificed by humans.

The phrase *natural capital* is used to refer to those features of nature that are utilised or have the potential to be utilised in human social and economic systems. This is an especially difficult concept, due partly to the uncertainty about potential future uses of natural features. However, because of this uncertainty, and because features of nature that are utilised as direct inputs to economic processes are generally dependent on others that are not currently considered to be inputs to economic processes, a broad definition of natural capital is preferable to a narrow one.

Natural capital is subdivided into *critical* and *non-critical* natural capital. The term critical is applied to irreplaceable, irreparable, or particularly scarce elements of nature, and to those elements that are believed to play a particularly significant role in maintaining some important or desirable ecological state. The term non-critical is applied to both renewable and non-renewable elements that are currently abundant. It is also generally restricted to those elements that, it is believed, can be utilised with relatively few ecological consequences.

Definition of natural capital

Natural capital refers to those features of nature, such as minerals, biological yield or pollution absorption capacity, that are directly or indirectly utilised or are potentially utilisable in human social and economic systems.

The nature of this universe is such that everything, ultimately, tends to maximum entropy. This means that life, which is a state of low entropy, must continuously capture negative entropy or cease to exist. Life on Earth is made possible by the fact that energy is given off by the sun, some of which is captured. Plants capture and store solar energy directly, by rearranging molecules to a state of higher potential energy (lower energy), in a form that can then be accessed by herbivores, who may in turn represent a food source for further conversion. Humans can utilise this continuous flow of energy in a number of ways, by eating vegetables or by eating sheep that have eaten grass, for example, and thereby capturing some of the negative entropy for their own use.

Humans can also move materials up into a state of low entropy. Pure copper wire, for example, is in a state of lower entropy than copper-bearing ore. This is achieved by input of energy from other sources (via the operations of mining, refining, and machining).

Such processes always incur a net loss. The amount of energy present in the final form is always less than the amount of energy required to manufacture that form. This is because of conversion losses. When a carnivore eats a herbivore, for example, a typical conversion loss is of the order of 90 per cent. Conversion losses mean that the further away from the primary conversion of solar energy (usually referred to as being 'higher up the food chain') the smaller are the numbers that can be supported, which is why there are always more deer than tigers. However, it is the distribution of the entropy that is important. As long as entropy is kept low locally life can exist, even though entropy will be increasing overall.

The process of evolution has ensured that genes code for behaviour that will make each species more likely to solve its version of the basic entropy problem. This is usually expressed in terms of the needs to survive (maintain low entropy at individual level) and reproduce (maintain low entropy at species level). The contingent nature of this process has given rise to a wide variety of strategies for achieving these goals. In humans, these strategies are usually expressed as a small number of basic needs, for food, shelter, and opportunities to reproduce. Humans are also, to a unique degree, a cultural species, and have needs for a variety of other social interactions. These needs generate a minimum level of demand for resources, which will be determined by the size of the population and the average per capita demand. Other demands have been created by the extension of the human species into a wide variety of environments, by the development of complex and diverse modes of organisation, and by changes in relative cultural values.

The human species therefore now requires a wide range of resource inputs. This includes current biomass for food and fuel, fossilised biomass for fuel and chemicals, minerals for chemicals and construction materials, and so on. All of these inputs can be described as natural capital, as they form the material basis for all human activity.

A number of human activities, such as agriculture, involve converting energy from one form to another. As discussed earlier, such conversions incur lsses. Typical conversion losses are 20–1 in corn-fed cattle (converting grain to meat), 8–1 in grain-fed pigs (a similar conversion, but more efficient) and 60–1 in scrub-grazed cattle (converting vegetation with a relatively low energy value into meat).[52] Some of these conversion losses are probably incurred unneccesarily, as edible and palatable forms of protein are converted into other forms of protein. For example, some of the world's fish catch is converted to fishmeal, some of which is then fed to farmed fish for reconversion into fish flesh. Various inefficient processes like this exist because of cultural preferences.

Natural capital is a fluid concept. Some types of natural capital (such as cereals) only exist in the forms (cultivars) and to the extent that they do as a result of direct and continuous human input. Some types of natural capital, such as transgenic mutations, may not even have existed previously in nature. However, the original materials (grasses and genes, respectively) are natural in origin.

It is useful to retain a distinction between natural capital and artificial capital, however, where natural capital refers to materials and processes in a relatively unaltered form, while artificial capital refers to materials and processes in a form that has been artificed by humans. In some cases this will be a clear distinction, in others it will not. In those cases where there is no clear distinction, it may still be useful to retain a conceptual division at the point at which the largest input of energy is required to change the form of the capital, as this is significant in thermodynamic terms and hence in terms of human ability to continue using that form of capital.

There are, of course, elements, substances, and organisms on the planet which have not yet been utilised directly by humans. These too should be included in any definition of natural capital, for the following reasons.

❑ *Ecological interdependence.* Many of the features of nature that are directly utilised in economic processes are dependent on features of nature that are not directly utilised. For example, current biomass depends on an ecological infrastructure. This means that it is necessary to include both those elements that *constitute* flows into human systems and those elements that *enable* flows. The sustainable harvest rate of a given species of fish, for example, will depend on the maintenance of the complex web of relationships that constitutes the ecology of that species.

This is also true of many forms of pollution absorption capacity. For example, the use of fossil fuels entails the release of carbon. The rate at which these fuels can be used without adverse consequence, such as global warming and so on, will be determined by the rate at which the planet's biological and geochemical carbon cycle processes can absorb the excess carbon. In this case, there are elements of nature that directly absorb this pollution, such as the growing vegetation that absorbs carbon, for example, and there are elements of nature that enable that absorption, such as the topsoil and bacterial action which support the vegetation.

❑ *Future use patterns.* It is impossible to anticipate all future needs or use patterns. Consider the following examples.
   – Technology might enable the use of a previously untapped or uneconomic resource. Technology might also make it possible to use certain resources for higher value purposes (see below).
   – Energy limits may make the extraction of certain energy resources non-viable. The energy cost of oil extraction may eventually make the utilisation of oil as a fuel uneconomic, although it would probably still be economically viable at that point to extract the oil for higher-value uses, such as a source of lubricants and chemicals.
   – New information about the global ecology may indicate that the

human species is dependent on the maintenance of some natural feature to an extent that was not previously understood.

❑ *Patterns of indirect usage.* It is not even possible to be sure that we are aware of all current patterns of natural capital usage. While catching fish is a fairly obvious way of drawing on the fish stock, it is less obvious that using an underarm deodorant may also draw on the fish stock (via the CFCs in the aerosol propellant, ozone depletion, increased UV-B penetration, and impact on marine biota – see p 66).

These points suggest that the only candidates for non-inclusion in the category of natural capital should be those features of nature that demonstrably have no current or potential utility whatever. As it is impossible to prove a lack of potential future utility, we conclude that all features of this planet must ultimately be included in the category of natural capital. In many instances, however, it is likely to be useful to maintain a distinction between direct and indirect uses of natural capital, and between those elements of natural capital that constitute flows and those elements that enable flows.

In summary, then, natural capital includes those features of nature, such as minerals, biological yield, and pollution absorption capacity, that are directly or indirectly utilised or are potentially utilisable in human social and economic systems. This ultimately includes, directly or indirectly, most or all of nature.

Natural capital and artificial capital have certain important differences. Because of the nature of ecologies, biological natural capital will be subject to irreversible change at certain system thresholds (see p 31). This means that it is likely that there will be critical minimum stocks of certain types of natural capital, which should not be depleted if a particular ecological balance is to be maintained.

This particular feature of natural capital is particularly significant in situations where it is necessary to estimate the risks associated with a particular development. The costs, in the most extreme case, of a major change in the global ecological balance may be effectively infinite. This would have the effect of heavily skewing a cost–benefit analysis, thereby resulting in a more risk-averse strategy.

Nature, of course, also possesses other values for humans. Some of these are based on moral and philosophical positions, such as a belief that other forms of life also have intrinsic value, while others are based on aesthetic judgements and feelings. As these intangible values clearly do affect human behaviour, it is important to take them into consideration. It would be possible, for example, though with some practical difficulties, to include a broad definition of landscape quality in the concept of natural capital. This could include some consideration of the natural beauty, the historical sense of the land, religious and spiritual values, and so on. The practical difficulties result from the fact that, for example, religious values are not held in common and aesthetic values often differ widely, and neither will necessarily be closely related to other important factors, such as the ecological condition of the land.

Categories of natural capital

Natural capital can be categorised in different ways. For many purposes, the important distinctions are whether the source is renewable or non-renewable, and whether the capital is in the form of material or energy.

The significant renewable materials are those that are biological in origin, such as timber, grains and animal products. The renewable energy sources are solar, gravitational, geothermal, and biological in origin.

Non-renewable materials are elements and compounds of natural origin, in the form of minerals, metals, gases and so on. The significant non-renewable energy sources are the fossil fuels and fissile materials.

Natural capital depletion

It is possible to draw on natural capital in different ways, some of which allow the natural capital to be replenished and others of which do not. The latter would be unsustainable if pursued for a sufficient time. In the case of an extremely abundant natural resource, however, this sufficient time may exceed the likely lifespan of the human species.

The depletion of natural capital in the form of materials should be considered separately from the depletion of natural capital in the form of energy. These two areas must in turn be subdivided into renewable and non-renewable sources. A further distinction has to be made between critical and non-critical natural capital.

Materials

One of the key distinctions is that many renewable material resources are biological in origin and replenish themselves, while many non-renewable materials are not biological in origin, and do not replenish themselves.

In practice, the distinction between renewables and non-renewables is partly arbitrary and based on timescale. For example, topsoil can be replaced as new topsoil is created by weathering of parent rock and by biological action. However, topsoil regeneration usually occurs at a slow rate that places it outwith a timescale meaningful to humans (see p 68), so it is not normally considered to be a renewable resource. Timber, however, is considered to be a renewable resource, even though hardwoods may not reach their full potential value for one or more centuries.

The ozone layer is a renewable resource, because ozone is continuously created and destroyed in the atmosphere (see p 64). However, the protection afforded by the ozone layer can be thought of as a resource which is utilised in situ by being left intact. This amounts to treating stratospheric ozone as a critical non-renewable resource. Its use as a sink for the temporary disposal of chlorine effectively uses up some of its yield as a renewable resource, thus reducing the protection available. If, however, chlorine deposition is halted the layer maintains the capacity to recover once the chlorine leaves the stratosphere.

Fossil fuels are an important case. Carbon is captured by trees and other plants from the atmosphere. With sufficient time, heat, and pressure this biomass is converted into coal, oil or gas. When these fossilised hydrocarbons are burnt some of the energy originally captured from the sun during the late Palaeozoic, some 590 million years ago, is released. Much of the carbon itself is also released back to the atmosphere. The release of energy can only happen once, albeit possibly in several steps. This carbon could in principle be recaptured and refossilised at some point in the future, but this cycle will be millions of years long. This means that fossil fuels are effectively non-renewable. So the concept of renewable should be understood to refer to resources that are capable, at least in principle, of self-replenishment in a period of time that human society can encompass.

The renewal of all renewable biological capital depends ultimately on the input of 'free' energy from the sun. Free, in this context, means from outwith the system. This input makes it possible for biological activity to occur. Biological activity captures some of this energy by re-arranging molecules to a state of higher potential energy. This chemical energy is then available for future conversion. So, as discussed earlier, plants capture solar energy and store it in a form that can then be accessed by herbivores, who may in turn represent an energy source for carnivores and omnivores.

So biochemical processes of life on this planet, utilising 'free' solar energy directly, will continuously make energy available in assimilable forms, will continuously capture molecules from air, water, soil, and other life forms, and will continuously rearrange these molecules into more complex (and potentially more useful) forms. In this way biological activity uses solar energy to reverse the normal increase in entropy by arranging matter into more highly ordered states.

Non-biological capital is not usually renewable. Although most of the materials used by humans are not destroyed (at an atomic level) by that use, they may be permanently transformed into a state where humans can make no further use of them. This is typically because human use involves capturing some of the low entropy of the capital, thereby leaving it in a state of increased entropy. The capital could not then be re-used without an input of energy to restore its lower entropy. The energy input would have to be greater than the energy that was captured in the first conversion, because of the two sets of conversion losses.

Minerals and metals may be mined, refined, and concentrated into utilisable form, but with few exceptions (mostly via nuclear technology) no new atoms of the elements will be created on this planet. In general, any human developments will for the foreseeable future be based on the existing stock.

As the richest and most accessible reserves of minerals and metals are normally quarried first, there is a common long-term tendency for costs of extraction and processing to increase as poorer or less accessible reserves are brought into use. It is now considered to be worthwhile to extract oil from relatively high-cost environments, such as the North Sea and Alaska, and to develop proposals to drill in Antarctica. The trend can also be seen in the pattern of exploration and development in the North Sea, where the era of the large fields (such as the Brent fields) is ending, and the focus is moving to the development of techniques that make it possible to extract smaller and smaller pockets of oil economically.

If a given mineral or metal, albeit in a form with higher utility, is diffused through an economic system without a facility to capture and re-use it, this has the effect of dissipating the material by dispersing it finely across the surface of the planet. There might then be a diminishing number of sources of the mineral or metal that can be mined economically. For many materials this holds true even if the price of the material rises steeply, for reasons that are discussed on p 99.

One possible consequence of continued use of finite resources is that, at some future point, human waste dumps may become the richest remaining reserves of certain materials. This cannot indefinitely postpone depletion beyond the point of economically-viable recovery, as only some fraction of the total reserve will have ended up in known dump sites, and as the energy costs of extraction may be high.

Renewable resources can themselves be utilised at sustainable or unsustainable rates. Any biologically generated renewable resource has a certain potential surplus that can be utilised.

To understand this, it is necessary to understand the various evolutionary strategies. Reproductive strategies lie on a continuum, with extreme 'r-selection' strategies at one end and extreme 'K-selection' strategies at the other. Briefly, r-selection species tend to be small, and to have large numbers of offspring which afford them relatively little protection once they are born, whereas K-selection species tend to be large, and to have small numbers of offspring, but then support and protect them when they are young.

This means that r-selection species must reproduce at rates far above those that would be required for simple replacement. This strategy generally incurs (but in effect allows for) high rates of losses. It is these 'tolerable' losses that form much of the utilisable harvest.

As species (of either r-selection or K-selection type) colonise an area, the general trend is from a situation where pioneer species predominate to one termed the climax situation. The climax situation is reached when the ecosystem is, in effect, 'full'. An econiche could be filled, typically, by relatively large numbers of an r-selection species, or relatively small numbers of a K-selection species. Biological diversity is one result of this process of evolutionary competition, as the available 'room' is further divided and subdivided.

The cultivation of crops and the domestication of animals represent particular sources of selection pressure. Essentially, the management of plant and animal populations consists of a process of simplification of the ecology concerned by reducing or eliminating competition. Selected r-selection plants, for example, will then expand into this new econiche, as other plants are cleared away and other predators destroyed. This process creates the maximum possible number of the chosen species, which are then harvested for human consumption. As human influence has spread, an increasing percentage of the planet's biologically-captured energy has been diverted in this way for human consumption.

This process of simplification invariably incurs certain costs. Large inputs of energy and other resources are required to bring about the desired effect. The expansion of the human species into a large range of very different ecosystems has been made possible by the availability of large quantities of energy derived from fossil fuels.

Furthermore, extensive cultivation may, by excessive reduction of competition, eventually have the effect of reducing the total genetic diversity. A successful alien intrusion, by definition, takes up some of the 'room' in an ecosystem.

Further expansion of the human species would be possible, depending on the availability of sufficient quantities of acceptably low-cost energy, but would inevitably incur other costs in terms of further reduction of total biological diversity.[53]

In any discussion of sustainable yield, therefore, it is important to bear in mind that most forms of biological reproduction that are currently harvested have reached their current distribution and gross productivity as a result of extensive human intervention. Other forms of biological reproduction that are now being harvested without prior human cultivation (such as the logging of virgin forest, or the development of new fisheries) represent an initial or pioneering expansion of the human species into a previously untapped ecosystem.

Given these caveats, the general principle is that any biologically generated resource can be harvested at rates that are within or above its surplus yield. To harvest above this rate will, over a sufficient period of time, render a renewable resource effectively non-renewable. To harvest within this rate is sustainable for the foreseeable future, that is, as long as the necessary 'free' inputs of biological action and solar energy continue to be available (see also p 88).

Energy

Energy can be derived from renewable and non-renewable resources. The term renewable is usually applied to energy in a wider sense than that in which it is applied to materials.

For example, energy can be derived from biomass. Fossilised biomass, such as coal or oil, is a non-renewable source of energy. Current biomass, such as timber, is a renewable source of energy, provided that the volume of biomass extracted does not exceed the surplus capacity (in the sense of its ability to maintain at least replacement rates of reproduction) of the ecosystem.

However, energy can also be derived from solar flux and the gravitational attraction of the moon. Since these sources are in effect outwith the Earth system and are likely to be indefinitely available, these applications are also termed renewable. The costs of using them are largely limited to the resource costs of the construction and maintenance of the necessary technologies. There are few known environmental problems with their use other than effects such as visual intrusion and purely local ecological impacts. The total sustainable yield from these sources is far greater than any likely human demand, so the only limiting factor is the efficiency with which these sources can be tapped. This is a good example of how appropriate technological development could transform the effective resource base by permitting a resource to be utilised in a more effective manner.

The primary renewable energy sources are solar, gravitational, geothermal and biological. The applications include the following:

❑ photovoltaic cells and solar panels, which capture solar energy flux directly;
❑ inland hydro-powered generating stations, which utilise the hydrological cycle, which in turn is driven by solar input;
❑ tidal barrages, which utilise the tides, which are in turn driven by the gravitational attraction of the moon;
❑ on- and off-shore wave power generators, which utilise waves, which are in turn driven by winds, which are driven by the temperature gradient of the planet, which in turn is a result of solar input; and
❑ aerogenerators, which capture energy from winds.

Depletion of natural capital: conclusions

Any use of a non-renewable resource can be classed as unsustainable unless one of the following conditions is met:

❑ The resource is so abundant, and is present in forms that can be accessed for low energy and resource costs, that there is no possibility of ever experiencing any scarcity within any meaningful timeframe.
❑ There are mechanisms to recover and recycle the resource within the economy with a high degree of efficiency.

Use of a renewable resource is in principle indefinitely sustainable, provided that the harvest taken does not exceed the surplus yield capacity of the resource. If, however, the harvest itself is dependent on continuous inputs, such as fertilisers and pesticides, the sustainability of the harvest will be determined by the sustainability of these inputs (see p 68). Similarly, if the harvest incurs other extraneous costs, such as the loss of biodiversity or soil erosion, the sustainability of the harvest will depend on whether these costs remain tolerable.

Critical natural capital

For practical purposes, it is probably necessary to make an additional distinction between critical and non-critical forms of natural capital. The term 'critical' is generally applied to elements of nature that are both irreplaceable or irreparable and currently scarce. The term is also applied to those elements whose loss would cause significant further loss. This would include, for example, a species at the base of an extensive food chain. It would also include the minimum extent of a habitat or range of habitats necessary to maintain a particular ecological balance, or any element that plays a significant role in the maintenance of global ecological conditions. A distinction should be made between local criticality (when the loss of an element would make a significant difference to a particular region) and global criticality (when the loss of an element – such as the ozone layer – would have genuinely world-wide consequences).

The term 'non-critical' is generally applied to those elements that are renewable and currently abundant, and to non-renewable elements that are present in such abundance that there is little likelihood of exhaustion within any meaningful timescale. It is also generally restricted to those elements that are currently believed to be less ecologically sensitive, and which can be utilised with relatively few environmental consequences.

The term 'critical' is always relative. Something that is critical for one species or person may not be critical for another. An event that eliminated the human species but left anaerobic life forms otherwise unaffected might not appear to be a loss of critical natural capital from the point of view of the anaerobic life forms. In this case, criticality refers to the potential implication for human society of the loss of some particular form of natural capital.

It is important to note that it would be possible for some elements of natural capital to be critical without necessarily being scarce at the time. Given the uncertain knowledge of possible ecological thresholds, it might be possible to approach a threshold of system tolerance while the element of critical natural capital being depleted was still apparently relatively abundant, and therefore cheap (in economic terms). This is one reason why we cannot rely on economic signals to warn of possible environmental damage (see p 31 and p 107).

There is a continuum between critical and non-critical, and a given resource may become more or less critical. A resource would become more critical as it grew scarce, particularly if it were believed that further use would incur some risk of crossing an environmental threshold and triggering wider ecological consequences. Such perceptions of criticality could be inaccurate, of course, as they would depend on the estimates of the damage which would result from further depletion. Thus it would be possible, for example, to move a renewable resource from the non-critical to the critical category by over-exploiting it, that is, by extracting more than the maximum sustainable harvestable yield over a long enough period of time to diminish the capacity of the resource to recover. On the other hand, good conservation could have the effect of allowing stocks to replenish and thereby moving that resource into a less critical category. In other words, the distinction between critical and non-critical is dynamic, and depends at least partly on the pattern of use.

With non-renewable resources different criteria are likely to apply. Any use of a non-renewable resource is not, per se, indefinitely sustainable.

This might not be significant, however, if the resource were so abundant that there was no possibility of exhaustion within any meaningful timescale, but it might be highly significant (and hence have the effect of placing that resource in the critical category) if that resource were not abundant. Similar considerations apply to the use of any resource where the use itself entails some pollution. This might not be significant if the pollution caused were negligible in terms of its ecological impact or in relation to the limits of the planet's ability to absorb the pollution, but it might be highly significant (and therefore have the effect of placing that resource in the critical category) if the effects of the pollution were found to be significant in ecological terms.

In general, therefore, the degree of criticality can vary. Forms of natural capital should be regarded, for management purposes, as becoming more critical as they become more scarce or as estimates of the scale of the potential ecological implications of their further depletion are revised upwards.

The priority attached to decreasing demand for a particular resource should therefore depend on whether that resource is critical or non-critical, with higher priority being attached to critical resources. A low-risk strategy would be to ensure that reduction in demand for critical natural capital was achieved primarily either by developing alternative means of satisfying the relevant human needs and aspirations, or by developing means and technologies to utilise non-critical natural capital sources as alternatives.

Of course, it is often possible to improve the efficiency with which natural capital can be converted into artificial capital (see p 91), so that less natural capital is needed in order to supply the same amount of artificial capital. With critical natural capital, however, improvements in the conversion efficiency, although these would be generally desirable, might not be sufficient. Any improvements in the conversion efficiency would have to be both developed and implemented faster than the rate at which the natural capital concerned was becoming more critical.

The low-risk strategy would therefore be to aim for low or zero utilisation of critical capital, either on an indefinite basis or until such time, in those cases where this is applicable, as critical capital finally recovers to the point where it is non-critical.

Managing natural capital

Daly has suggested three rules for the management of natural capital.[54]

1 For a renewable resource, the sustainable rate of use can be no greater than the rate of regeneration.
2 For a non-renewable resource, the sustainable rate of use can be no greater than the rate at which a renewable resource can be substituted for it. This is in turn subject to the constraint above on the rate at which the renewable resource can be used.
3 For pollution absorption capacity, the sustainable rate of use (that is, the rate of emissions of pollution) can be no greater than the rate at

which that pollutant can be processed or absorbed by the environment.

These would be sensible rules, and have the additional advantage of being simple and clear. However, as with most such principles and rules, they would not be simple to apply in practice. Consider the following examples:

1 Complex calculations would be involved in estimating the rate of regeneration for some renewable resources. This is because, in some cases, the rate will depend on a range of both internal and external factors. The rate of regeneration of a species of fish, for example, will depend partly on internal factors, such as the reproductive cycle and the demography of the existing stock, and partly on external factors, such as the rate at which predator, prey, and competitor species are being fished. It will therefore be necessary to define and model a range of external variables to calculate an approximate regeneration rate for any one factor.
2 With a non-renewable resource, the rate of allowable use will depend on calculations as to what would form an adequate substitute, and whether such substitutes could and are being developed in time to support those economic functions that currently depend on the use of the non-renewable resource.
3 Similarly complex calculations would be involved in estimating the rate at which pollution absorption capacity is being consumed. The response of many ecological absorption systems is non-linear, so that they behave differently at different rates and aggregate volumes of input.

Carrying capacity

The concept of carrying capacity is fundamental to an understanding of the relationship between humans and the natural environment. Carrying capacity refers to the population that a given ecology can support. The main factors in determining carrying capacity are levels of population, patterns of resource demand, environmental yield potential and resource flows, and environmental absorption capacity and impacts. The interaction of these factors determines the long-term viability of development options.

It is important to note that it is the aggregate demand of a population that will ultimately be limited by carrying capacity. Aggregate demand may be computed as the average per capita demand (in terms of consumption and emissions) multiplied by the number of people.[55] As the number of people increases, therefore, the average demand must be reduced unless there remains a margin of yield (which means that the current demand is within the carrying capacity limit), or unless the yield potential can be increased at the same time.

Yield potential can be enhanced (to a degree) or diminished (probably with less constraint). In the short term, increases in yield can be achieved by drawing on natural capital reserves. In agriculture, for example, crop yields can be boosted with inputs of fertiliser, pesticides, fungicides and mechanical cultivation techniques. These inputs currently depend on chemicals and energy derived from fossilised hydrocarbons, which are a non-renewable resource (see p 81). These increases cannot therefore be indefinitely achieved in this manner (see p 84).

Long-term increases in the overall efficiency of production are possible. Methods that reduce the dependence on the use of non-renewable resources are more likely to be indefinitely viable. Such methods may include the following:

❑ broadening human dietary preference (see p 72);
❑ shifting the main source of human nutrition further towards the primary conversion end (see p 81);
❑ developing cultivars and variants of plants and animals that are more efficient at converting renewable inputs or are able to utilise low-grade and waste sources of nutrition. Transgenics may have a certain potential in this area in future. However, the release of transgenic organisms into the environment is invariably dangerous. Moreover, as a general principle it would be better to address any political, economic or cultural problems that might be causing food shortages directly rather than seeking to ameliorate their effects with a technofix.

If human demands were ever to approach the limits imposed by the carrying capacity of the planet, it is likely that other environmental and related costs would start to rise steeply. For example, it might in principle be possible to utilise the planet to its full human carrying capacity by using every part for urban and industrial sites, mines, and grain- and grass-growing farmland. However, this would represent a drastically simplified global ecology. Simple ecologies can be less stable and more prone to oscillation than complex ones (see p 31). The likelihood of global ecological catastrophe might therefore be greater than with a complex and diverse global ecology. The quality of life is also likely to diminish as carrying capacity is approached.

If human society is to achieve a sustainable point of balance, the need to maintain the necessary global ecological stability (with acceptable margins of risk) is likely to define limits to the maximisation of production. This point, given current indications of environmental stress, is likely to be considerably below the theoretical human carrying capacity.

A similar problem arises with the calculation of the maximum sustainable harvestable yield of any particular species. When a species is expanding to 'fill' an econiche, the growth curve will typically form an 'S' shape (see p 37). The steepest gradient on the curve is where the population is growing fastest. This is sometimes referred to as the maximum sustainable yield point, where the greatest harvest can be taken. This is not usually completely accurate, as the ecosystem itself is likely to have different characteristics at that stage. This is so precisely because the species in question has not stabilised, and so the ecosystem is not at a climax situation. The true maximum sustainable yield rate could only be calculated with a multi-species analysis, and the necessary information is often lacking (see p 84).

Carrying capacity can be calculated at regional as well as global levels. Each country has a certain carrying capacity. This is usually obscured by patterns of world trade, which have the effect of transferring carrying capacity. A flow of resources from natural capital-rich to natural capital-poor countries will allow the natural capital-poor countries to maintain a higher level of population or higher levels of consumption than they could otherwise achieve. It may of course be that national carrying capacities could not all be simultaneously reached without exceeding the global carrying capacity.

There are some circumstances under which transfers of carrying capacity would no longer be viable, such as in the following cases:

❑ if it became impossible for the natural capital-exporting country to maintain the outward flow of resources. This might happen if, for example, that country experienced population growth or demand growth to the point where the margin for export was absorbed. With countries that combined population growth with resource depletion, the margin might even become negative as the country became an importer rather than an exporter; or

❑ if the maintenance of the flow of resources itself incurred high energy or resource costs, and such costs rose to the point where the operation became marginal. If the transfer of resources depends on the use of energy-intensive forms of transport, for example, the viability of the transfer operation depends on the continued availability of sufficiently cheap energy.

International trade

With a natural resource-based economy it is theoretically possible to calculate the sustainable yield rates of renewable resources and the life expectancy of non-renewable resources. However, it is less clear what sustainability would mean for an advanced industrial or post-industrial manufacturing and trading nation such as the UK. Many such developed countries sell manufactured goods and services, and buy in natural capital in the form of raw materials (see p 88), ranging from minerals to food. Such nations will, by definition, experience significant flows of both natural and artificial capital across their borders.

The current economic organisation of such nations depends on the existence of natural capital-exporting nations. This is not, therefore, a mode of economic existence that is available simultaneously to all nations.

At the global level, such flows become internal and can be aggregated. At the national level, the issue is more complex. It is necessary to disaggregate the types of capital involved. For example, outward flows of non-renewable resources, such as minerals, are not indefinitely sustainable (although this may not necessarily impose many practical restrictions in the case of very abundant resources), and the same applies to outward flows of renewable resources harvested at biologically unsustainable rates. These should be distinguished from outward flows of renewable resources harvested at sustainable rates, which do not suffer from the same limitations.

On the other hand, inward flows of natural capital may constitute a form of 'imported sustainability', which would not become apparent until assessed at global level. Japan, for example, has met certain domestic needs largely with imported natural capital. An instance of this is that Japan has protected its own forests well, but has done so partly by becoming the largest timber importer in the world.

Similarly, the net impact on global sustainability of imports and exports of artificial capital will depend in part on the form of the artificial capital and the form of the natural capital from which it is produced (see p 96 and p 129).

At a global level, any move towards greater efficiency of energy and resource use and conversion would probably be helpful. It is important to add that this would also be a sensible strategy from the narrower perspective of national self-interest. Improvements in the efficiency with which natural capital can be converted into artificial capital, and in the consequent productivity of the national resource and human capital, may well confer competitive advantages in the long-term, because there are factors (discussed on p 99) which are likely to lead to long-term increases in real prices for the resource inputs. If this does prove true, it would be essential for developed countries such as the UK to improve their energy and resource productivity and thereby to reduce their costs in order to continue to compete in world markets.

Artificial capital

Artificial capital refers to the tangible and intangible goods and services that society produces, the tokens and media of capital exchange, and other factors of production, such as a knowledge base.

Converting natural to artificial capital

Advances in material standards of living depend on three factors – knowledge, energy, and materials. Materials and fossilised forms of energy can be thought of as natural capital. Human social and economic systems depend ultimately on the conversion of natural into artificial capital, that is, into forms that humans can utilise directly. Artificial capital refers to the goods that society produces, in contrast to the 'free goods' that nature provides.

There are several important points about this conversion process.

❑ *Conversion cannot be 100 per cent efficient.* The process of conversion is usually thought of as 'adding value', because inputs such as human knowledge and labour have increased the utility of the material. This increase in utility should not be confused with the fact that conversion always incurs an actual cost in terms of energy, materials or both.

❏ *Artificial capital (with the exception of knowledge) depreciates.* This means that it degrades, loses utility, and is eventually lost from the economic system. If a motorway, for example, were to crumble to dust, the same amount of sand, lime, iron and so on would be physically present but it would be in a form that would be less useful. Energy would have been dissipated and would therefore also be in a less utilisable form, but energy itself cannot be destroyed or lost.

To summarise these two points, conversion always has a real cost, and anything converted has a finite life span and must be replaced at some point.

The conversion process can be relatively short, as in a hunter–gatherer society. In an industrial or post-industrial urban society with a diversified economy, however, the conversion process is typically lengthy and complex.

To build a house in the UK, for example, may require that clay and other minerals be quarried in the south of England, transported to kilns in the north of England, batched and baked with energy that may be derived from uranium brought from Africa or coal shipped from China, turned into bricks, and then transported to rendezvous with hardwoods from South America, softwoods from Scotland, tiles from Italy, slates from Wales, wiring containing copper from Australia sheathed in plastic that was once oil under Kuwait, and so on.

As conversion losses are incurred at every conversion point, multiple conversions incur multiple losses. Although an efficient multiple conversion can be more economical than an inefficient single conversion, in practice complex or long-chain conversions of natural to artificial capital tend to be relatively costly in terms of the energy that must be used or material wasted. From this in turn it will be apparent that the economic organisation of the developed nations has been made possible by readily available cheap energy.

This situation would become unsustainable if any of the following circumstances were to arise:

❏ if the availability of fossil fuels were to decline, and if other sources of energy had not been not developed as substitutes;
❏ if the growth in energy demand outstripped increases in energy conversion (ie generating) capacity and improvements in energy efficiency; or
❏ if the energy conversion process generated pollution at levels or rates that proved to be in excess of the planet's ability to absorb them.

The idea of efficiency is very important. It has to be applied over the operation as a whole. Take the house example above. Imagine that a building conglomerate that manufactures building materials and erects buildings decides to improve the efficiency with which it converts natural to artificial capital. One option might be to ensure that the building products were manufactured with maximum energy efficiency and minimum wastage. Raising the energy efficiency of the brick kilns by, say, 20 per cent, might represent a large margin at that stage of the process. However, the energy cost of the materials only represents a part of the total energy and material cost of the final product, the house. So the percentage saving at that final product stage will not be as great as the margin that was achieved at the earlier stages. Compare that with another option, a decision to source most of the building materials locally. The effect would be to reduce the energy needed for transport across most of the range of inputs. The reduction in total energy cost at the final product stage might prove, in this hypothetical example, to be greater than the sum total of marginal improvements in the efficiency with which the building materials were manufactured.

This means that options for improving the efficiency with which energy and resources are used, such as improving the efficiency of specific conversion steps or eliminating conversion stages, should be compared in terms of the final potential saving. This analysis should be done in terms of the complete life-cycle of the product (sometimes known as 'cradle to grave'), as a product that is relatively cheap to construct may be relatively expensive in terms of running costs during its lifetime. This is, in fact, likely to be true in the house example given above.

Any attempt to improve the overall efficiency with which natural capital is converted to artificial capital will, in practice, be complicated by the fact that construction costs and running costs are often borne by different people or organisations. In order to ensure that the appropriate level of capital is invested at the point of construction, so that running costs are reduced to the point that gives the greatest possible increase in efficiency overall, it will have to be possible to pass on some or all of any increase in the costs of construction to the consumer. Any policy to improve the efficiency with which natural capital is converted to artificial capital, therefore, would probably have to make provision for measures to encourage people to preferentially purchase goods that were initially more expensive but of higher quality, so that they had longer lifespans and lower running and repair costs (see p 96 and p 176).

Net artificial capital growth

Net artificial capital growth is a highly aggregated concept. Artificial capital itself comprises both goods and services (although there are currently more conceptual difficulties with intangibles), capital that is reinvested and current consumption.

Growth in the total stock of artificial capital must be understood in terms of the relationship between production and patterns of consumption. Real or 'net' growth in artificial capital represents what is left after the growth in aggregate demand (which in turn is determined by the population growth multiplied by any change in the per capita demand) is deducted from any gross artificial capital growth.

Society currently has a net artificial capital growth demand.[56] This is for the following reasons:

❑ Artificial capital degenerates and depreciates, and is in other ways lost from the system. These losses must therefore be made good if the artificial capital stock is to be maintained. This means that there will always be a certain minimum level of demand for artificial capital.
❑ Rising living standards have generally been correlated, to date, with increased demands for natural capital.
❑ The human population is currently increasing, and this will generate increased demand for artificial capital.

If global ecological pressures now make it necessary to introduce cleaner technology throughout industry, for example, this would immediately create a large investment and artificial capital deficit. It might not, under some circumstances, be possible to meet the additional demand for artificial capital by converting more natural capital, as this might exacerbate the problem.

This suggests that, if it proves necessary to restructure the productive base of society, resources might have to be diverted from current consumption into investment.

This raises a number of complex issues. For example, societies at different stages of development and with varying degrees of wealth demonstrate different proportions of investment and consumption. In general, the more extreme poverty becomes, the more investment – even essential investment – has to be sacrificed to meet current demands in a way that is analogous to eating one's seed-corn. This reduces the extent to which the necessary resources could be diverted into reconstruction.

More fundamentally, in certain critical respects, there is no absolute distinction between consumption and investment. The difference between the two is essentially the duration of the time taken to consume the resources. This means that consumption and investment should really be regarded as points on a continuum. More precisely, consumption can be thought of as disaggregating into revenue expenditure (items consumed over the short term), capitalised revenue expenditure (revenue expenditure taken over a longer period), and capital expenditure (items consumed over the long term).

In other words, whether for consumption or investment, the resources are still being consumed. A strategy of encouraging investment at the expense of consumption in order to release the resources necessary for a reconstruction of the productive base of society would therefore not be entirely sufficient. It would be necessary to also introduce other measures for distinguishing between forms of artificial capital in terms of likely social and environmental impact (see p 93). It also makes it essential to disaggregate the concept of artificial capital for some purposes and to distinguish between those commitments of artificial capital that form part of the background demand for resources, and those commitments that would form part of a reconstruction of the productive base of society (see p 100). The following points should also be noted:

❑ The existing world stock of artificial capital is very unevenly distributed. The need to generate additional artificial capital to meet the needs of the additional human population should not be confused with the question of the distribution of existing artificial capital. However, they are interrelated questions. The bulk of the increase of the human population from the current 5.2 billion to the projected 7.7 to 14.2 billion will occur predominantly in developing nations. These nations therefore have high net artificial capital growth requirements, while nations with a mature infrastructure and stable populations have low net artificial capital growth requirements.

❑ Aspirations play an important part in determining what forms and quantities of artificial capital are created. If a particular distribution of the artificial capital raised expectations and caused aspirations to grow faster than further artificial capital could be created, the artificial capital growth demand could actually increase as more artificial capital was created. This sort of problem might arise, for example, in a competitive and materialistic society with a growing economy, in which increasing aspirations of wealth created additional social pressure to create and consume artificial capital.

❑ The rate of population growth (see p 74) is such that the increases in demand are likely to outstrip the rate at which artificial capital can be produced. If measures are taken to stabilise the human population this would eventually reduce the net artificial capital growth require-ment to the level needed to compensate for depreciation and related losses, provided that per capita demand could be stabilised at the same time.

❑ The net artificial capital growth requirement indicates the level of growth necessary to maintain a given average level of consumption. If circumstances or changes in cultural expectations indicated that the average level of consumption could be reduced, this would reduce the net artificial capital growth requirement. Again, the reduction of the average level of consumption should not be confused with the distribution of levels of consumption. The average currently results from some people consuming a great deal and others consuming very little, and it is with the former that there is the greatest likelihood of finding a sufficient margin over subsistence levels to make a reduction both possible and worthwhile.

Any increase in population, of course, will tend to absorb some or all of any margin of resource supply over resource demand that might be created by a reduction in average levels of consumption and consequent reduction in the artificial capital growth requirement, making net reductions harder to achieve. Demand reductions should not be confused with reductions on the supply-side. Reductions in supply tend to cause serious dislocation. Factors such as resource exhaustion, biological depletion, and the reaching of energy or economic limits to extraction potential would eventually lead to a reduction in supply, and this might, with some time lag, lead to a reduction in demand, but the ecological and social cost would probably be very high.

❏ The net artificial capital growth potential is related to productivity. If economic productivity declines it becomes necessary to draw more heavily on existing artificial capital, assuming that existing average levels of consumption are to be maintained. If productivity goes down, therefore, the net artificial capital growth requirement would actually go up.

Artificial capital utilisation

Some of the forms of artificial capital that society produces have greater utility and value than others. Some, such as habit-forming drugs, junk food or 'snuff' videos arguably have negative utility.

For example, oil can be converted into plastic. The plastic could be used to form surgical prostheses which permanently improve lives, or it could be used to form shoddy plastic trinkets which are of less obvious benefit and which usually become a waste disposal problem shortly after purchase.

Decisions on the utilisation of resources, in a complex and differentiated market-based economy, are typically made at or near final product stage. The resource-demand pattern generated by these decisions then works back through the economic system against the direction of the flow of resources until it reaches the primary industries. The consideration of the uses to which resources are ultimately put is therefore more readily applied in the tertiary and possibly the secondary sectors than with primary industries. This means that it is important to consider regulating patterns of consumption as part of a transition to a more sustainable way of life.

Probably the optimal commitment of artificial capital, in the context of sustainability, is that which directly reduces demand for natural capital at the other end of the economic system. Investment in energy-efficient technology generally falls into this category. One example might be the construction of a modern energy- and resource-efficient transport system that gave long-term savings of energy and resources in excess of its energy and material construction costs.

This analysis must, as with the concept of efficiency (see p 94) be applied over the operation as a whole. The savings in the above example might still be marginal compared with, for example, the effect of changes in planning guidelines that would reduce the need for transport.

There are other, secondary, considerations, such as the contribution to social and economic welfare made by the commitment of artificial capital. It would be possible, for example, to accord a degree of priority to artificial capital utilisations in relation to their contribution to 'real' welfare, where real welfare includes only those forms that have positive utility (see p 176). This is a difficult question. There are likely to be a number of forms of artificial capital utilisation that are ambiguous or disputed. There are also moral, political, and philosophical dimensions to this question (see p 168).

Economic growth and the environment

We have indicated the need for growth in artificial capital (see p 93). This will in turn require a level of economic growth. However, if this economic growth causes further environmental damage, it will make it more difficult to achieve a sustainable state. It is therefore necessary to consider the complex relationship between economic growth and the environment in more detail.

Economic growth can be defined and measured in different ways, but essentially it refers to rising aggregate economic output or rising aggregate consumption. Consumption is usually a certain proportion of output, so these two measures tend to co-vary.

Both consumption and output are economic rather than physical measures. This means that growth in, for example, economic output does not necessarily require a growth in demand for materials and energy.

In practice, however, growth in some kinds of economic output clearly does require a growth in demand for materials and energy, and usually generates additional pollution. Suppose that some particular development increased economic output, but generated more pollution than could be cleaned up by spending all of the additional output on remedial works. It is easy to see how this situation might arise if, for example, the additional pollution caused an ecological threshold transition (see p 31) with an irreversible effect on the environment, but it is also possible that net costs are incurred in more routine economic developments as well. In such a situation, economic growth would inevitably cause environmental degradation.

However, should some other development increase economic output at the cost of zero or negative net growth in resource demand and pollutive emissions, then economic growth could occur at current or even reduced levels of environmental damage.

It is incorrect, therefore, to state either that all economic growth is bad for the environment, or that only a growing economy can afford to be environmentally clean. The effects of growth depend on its nature. This would include, for example, the pattern of resource demands and pollutive emissions that the growth generates.

In practice, the relationship between economic growth and environmental quality has been largely negative. However, the strength and direction of the relationship depends, to an extent, on the stage of development.

This should not be confused with the relationship between negative economic growth and environmental damage. These questions are related, but the pattern of cause and effect is not the same. In many parts of Africa, for example, pressure on the environment and rates of environmental destruction have increased as economic growth rates have become zero or negative. This is partly because poverty tends sharply to limit options, and create pressure to sacrifice long-term interests for immediate survival.

Although the negative relationship between economic growth and environmental quality does not exist solely in industrialised or industrialising nations, it is generally true that environmental quality tends to diminish particularly sharply as countries industrialise.

As countries become wealthier and develop a more service-based economy, the environment tends to become more of a priority. This might be thought to suggest that economic growth in a post-industrial economy with a large service sector is compatible with rising environmental quality. Unfortunately, the situation is not so simple. A wider analysis may reveal that the industrial base has been exported into less developed nations, via a pattern of overseas investments. As sustainability must ultimately be measured at a global level (see Chapter 9), this would not necessarily represent a solution, but simply a displacement of the environmental impact. Similarly, rapid and unsustainable rates of natural resource exploitation in non-industrial commodity-exporting economies are, in many cases, likely to be driven primarily by growth in demand from industrial and post-industrial nations which have either exhausted or are protecting their indigenous natural resource base.

It is also possible that environmental quality tends to drop especially sharply during the process of industrialisation not because there is any intrinsic reason for initial industrialisation to incur such heavy costs in environmental damage, but because this period usually coincides with major cultural transformation. Many institutions are transformed, replaced, or destroyed. It is possible, therefore, that cultural codes of checks and balances are lost at this stage. For example, religion has historically been the means for providing a structured world view, discussing intergenerational issues and producing codes and ideals of behaviour. This and similar structures are typically weakened by the growth of the more materialist philosophy that tends to accompany industrialisation.

These cultural sanctions may eventually be replaced by legal codes, including environmental property rights, legal and statutory environmental protection measures, and other mechanisms for securing redress and preventing environmental damage. There may well be a hiatus, however, as these legal codes rarely exist in countries at that particular historical stage of their development.

Thus the apparent levelling-off of rates of environmental damage in the mature industrial and post-industrial economies may be an artificial trend generated solely by inadequate controls at the outset. If this is the case, then the true long-term trend may prove to be one of slowing but continuing decline in environmental quality from the outset of the process of industrialisation. It would follow from this that economic growth and environmental improvement would co-occur only during the phase in which environmental externalities were being internalised with, for example, the development of major new industries based on cleaning up the operations of others.

It would also follow from this that unless there were a significant redirection of the economy, the primary trend of declining environmental quality would continue until reaching the ultimate ecological system limits, at which point economic growth would cease and human society would collapse.

Limits to growth

The standard model of economic growth is based on, among others, the following assumptions:

❑ There are no real limits to the expansion of artificial capital.
❑ Artificial capital can be an effective substitute for natural capital.
❑ Where artificial capital cannot be substituted for natural capital or where some particular resource is exhausted, a technological solution will overcome the scarcity.

One version of this standard model suggests that if no substitute for a resource exists, then there will be an automatic economic correction as the output of that resource reaches a plateau, and then declines. The relative scarcity of the resource will be reflected in the price, so that use will be progressively discouraged as the resource becomes scarcer.

A related argument suggests that if energy prices start to rise, energy efficiency will improve and energy consumption will decline, thereby automatically correcting this trend towards increasing resource consumption.

These assumptions and arguments may be incorrect, for the following reasons:

❑ Artificial capital can be thought of as a function of knowledge and resources, comprising energy and materials. Both energy and materials may be derived from renewable and non-renewable sources. The availability of these factors is limited in different ways, as follows:
  – Renewable energy is limited by the total accessible rate of supply.
  – Non-renewable energy is limited by the fixed stock available.
  – Renewable resources are limited by the maximum sustainable yield.
  – Non-renewable resources are limited by the fixed stock available.
  Knowledge, on the other hand, does not appear to be limited. It is possible that it is infinite. Certainly, the only input to artificial capital which appears to be indefinitely expandable is knowledge.
❑ Advances in the material standard of living have also depended on these two factors of knowledge and resources. Historically, energy and other resources have been cheap, so that knowledge has been the limiting factor. It is now arguable that under certain circumstances, resources could become the limiting factor. Certainly, in some sectors there is a general trend towards increased resource extraction costs. Between 1974 to 1984 the fraction of the GNP of the US accounted for by natural resource extraction rose from 4 per cent to 10 per cent. This suggests that more resources now have to be committed to make resources available, which would tend to reduce the margin of viability of such operations.
❑ Previous 'energy savings' have usually involved shifts in the kinds of

fuels used (like the current shift in the UK to use gas turbines to generate electricity, rather than coal). It is important to note that this process is directed at improving fuel efficiency rather than overall energy efficiency. When the indirect energy costs are included, net gains in overall energy efficiency tend to be smaller than the apparent gains. Such 'energy savings' may have given a misleading impression of the progress made to date.

There is, of course, scope for real improvements in energy efficiency. The most significant savings are often those that can be achieved by shifting between technologies in those sectors of the economy where this is possible, such as by moving traffic from road to rail-based systems, for example.

❏ There are, of course, ultimate limits to the available improvements in energy efficiency. Efficiency can never exceed 100 per cent. Very few conversions, in practice, approach this level.

To understand the more fundamental problems with the standard model of economic growth, it is necessary to examine the underlying assumptions. Two of the key assumptions concern *substitutability* and *technological advance*.

Limits to substitutability and technological advance

It would be possible to have a sustainable future in which economic output, consumption, and welfare grew without limit, provided that the following conditions could be met:

❏ All key resources had substitutes.
❏ It would always be technologically possible to make such substitutions.
❏ Technological progress would always enable substitutions before resources were used up.

Clearly, there is scope for improvement in the price ratio of inputs to outputs in the conversion of natural to artificial capital. Rising resource prices will tend to stimulate efforts to reduce costs, to search for substitutes, and to promote efficiency, so that fewer and less costly material resources would have to be input to produce a given unit of economic output. This would have the effect of maintaining or improving profits, provided either that the technical improvements (after deducting the costs of the development and implementation) improved the profit margin beyond the increase in resource costs, or that the reduction in costs was only partly fed through into any reduction in prices.

This economic response, however, is often confused with the question of the physical conversion of natural to artificial capital. It is important to note that these are separate questions. The more fundamental question is whether similar improvements in efficiency could be achieved in the actual physical conversion of natural to artificial capital. Of course, the laws of thermodynamics ensure that there is a maximum efficiency of conversion of resource inputs per unit of valued artificial capital output, beyond which it is impossible to go. In addition, all artificial capital (except capital in the form of knowledge) depreciates, and incurs further costs in natural capital for its maintenance. Furthermore, the development of new technology itself incurs costs, both direct and in terms of external adjustment costs.

More fundamentally, artificial capital (see p 96) cannot be assumed to be a substitute for natural capital on an equal basis. Some forms of artificial capital may indeed displace demand for natural capital. For example, energy invested in the manufacture of insulating materials can result in a net saving of energy. Other forms of artificial capital are unlikely to displace demand for natural capital. A third category of artificial capital is that which is likely to increase demand for natural capital. For example, the installation of an air-conditioning system will result in a further increase in energy demand. Similarly, the creation of artificial capital in the form of fashion items is likely to result only in further increases in demand.

So artificial capital can usefully be subdivided, in this context, into three categories:

❑ That which displaces demand for natural capital.
❑ That which does not affect demand for natural capital, other than in its own construction.
❑ That which increases demand for natural capital.

Of these three, only the first is likely to form a possible substitute for natural capital. The impact of the third category is likely to be negative, in that it will tend to increase the pressure on the natural capital stock and hence the need to develop further artificial capital substitutes.

Similarly, while it may be convenient for some purposes to use a concept of aggregated natural capital, in reality natural capital is not uniform. Ecological resources, for example, are typically multifunctional, and generally do not behave in a linear manner (see p 31 and p 86). This is especially important at ecological system thresholds, beyond which change may be irreversible.

The idea of being able to substitute artificial for natural capital is, therefore, only applicable within such thresholds, since irreversible change, by definition, does not allow further substitution to occur. Moreover, whereas natural capital is often multifunctional, artificial capital is typically unifunctional. While some item of artificial capital may adequately substitute for an element of natural capital in some specific context or function, it is unlikely that it will be able to replicate all functions of the original.

These constraints mean that there will be ultimate limits to the extent to which technology and artificial capital can substitute for natural capital, and hence to material growth.

These constraints also mean that as long as output and consumption depend on non-renewable resources, there must be continuous technical progress which permits the output per unit of resource input to rise without limit, otherwise utility and human welfare must inevitably decline as the resource runs out. As the laws of thermodynamics make it impossible for the output per unit of resource input to rise without limit, utility must decline in proportion to the extent to which output and consumption are dependent on non-renewable resources or renewable resources harvested at biologically unsustainable rates. In those cases where such resources also provide other environmental functions or amenities, and the use of these resources as economic inputs therefore entails a loss of environmental amenity, there will be an additional and concurrent loss of utility as the reserves of these resources decline.

It should also be noted that the fact that the technology, for example, to improve a conversion efficiency exists or could be developed does not necessarily mean that it will be developed or deployed. There usually have to be more pressing reasons and incentives to do so (see p 116).

Energy limits

The concept of substitutability was developed in an era of cheap fuel. In an era of cheap fuel, energy is not, in practice, a limiting factor. The question is whether energy limits might in future have an impact on economic systems.

Energy itself has a certain energy cost. Energy is required to extract and process fossil fuels, for example. The amount of energy made available as output from a given input of energy for exploration, extracting and processing is termed the energy output/input ratio.

There is evidence that the general extracted energy output/input ratio is gradually deteriorating. With oil, for example, a ratio of about 100–1 was obtained in the 1940's. By the 1970's the ratio had fallen to 23–1. The ratio for newly-discovered oil reserves is now about 8–1. It is currently economically viable to drill for oil in the North Sea and in Alaska, which are high-cost environments. The average extracted energy ratio, taken across all non-oil energy sources, is about 5–1 today.[57]

In spite of this deteriorating energy output/input ratio, the price of oil, for example, still appears relatively low. This is for a number of reasons. One is that technological innovation and improvement have increased the total viable reserves of fossil fuels by making it possible to extract economically from smaller or lower-grade deposits. Another is that competition between fuel sources, energy utilities, and energy producing and exporting countries has helped to keep prices low.

The Earth's non-renewable resources are clearly not infinite. Some economic analyses have assumed that the finite nature of resources and the diminishing returns obtained need not mean that extraction becomes non-viable, because scarcity will drive up the price to the point where it is again worthwhile to mine the resource. Energy analysis reveals, however, that a non-renewable energy resource may be no longer viable as an energy resource even while quantities remain in the ground. That point is reached when the energy cost of extraction and processing exceeds the energy content of the resource.

There are various exceptions to this general principle. For example, resources such as oil have more than one potential use. Oil can be burnt for heat, or turned into plastics. While it is impossible to profit by extracting oil for energy when the payback reaches 1–1, it may be viable to extract oil for other uses of a higher value, such as conversion to plastic.

A version of this principle also applies to many other non-renewable resources. The energy and other costs involved in mining and refining copper, for example, may not be very significant when these costs are low and relatively rich reserves of copper still remain. If the energy and other costs of extraction rise, and relatively poor reserves of copper must be brought into production, then the cost of the extraction as a proportion of the total value of the extracted copper must increase. This would make further production uneconomic at the point where the maximum value of the copper was less than the total cost of extracting and processing it. It is important to note, as with the oil example above, that with resources that have multiple uses it is the maximum use value that determines the point at which production becomes uneconomic (see p 86).

Similar arguments also apply to land use. For example, the total land area currently in agricultural production could be approximately doubled. However, the cost of doing so would rise as less suitable land (the equivalent of a poorer reserve) was brought into production, and as the cost of the associated energy inputs increased (see p 69). This would eventually limit further expansion, as the value of the output would at some point be exceeded by the costs of the inputs.

Resource uses and trade-offs

The issue of substitutability discussed on p 100–101 raises an important question as to whether every resource must be conserved, or whether trade-offs between different resources would make it easier to integrate economic and environmental demands.

Clearly, it would be necessary to have some sort of system for deciding on the relative importance of any resources to be traded, which could then be translated into values or weights which could be assigned to each resource, which would then determine the 'exchange rate' between resources. Something like this would have to be developed if this strategy were to be pursued.

There would be a number of political and technical difficulties involved, however, in developing this kind of resource exchange rate system. For example, there are several bases that one could use for measuring resource value. There is a case to be made for using market prices, but there are also cases to be made for using ecological values or physical measures (such as comparing between different kinds of fuel in terms of their energy content).

Then there are possible threshold effects (see p 31) and ecological irreversibility effects, which could impose limits on the kind and extent of the trade-offs that could be permitted. Unfortunately, it is not currently possible to assess the risks, uncertainties, and margins of error associated with such possible thresholds with any degree of certainty.

It is also likely that there will be, at any one time, technical limits to what substitutions can actually be made, and these will also influence what can be traded.

Finally, different resources are likely to have different values in different countries, or at different levels (local, national and global, for example), or to different generations (which is an even more intractable problem, as such values may be unknowable). These factors would immensely complicate the task of actually using resource values in any meaningful way. This tends to be even more difficult with resources that are not currently marketed, as there may be little or no indication as to what values might be assigned or as to how these might change.

One possible solution would be to assign weights to resources according to current economic productivity. It would then be possible to trade-off resources to maintain an economy's aggregate economically-effective resource base. However, this generates its own set of assessment and measurement problems.

For example, it might be possible to use a constant real price index for materials at the primary input stage.[58, 59] However, this would make no allowance for capital accumulation or technical progress, which might improve the efficiency with which natural capital can be converted into artificial capital and thereby reduce the quantity of materials needed to produce a unit of economic output.

It might also be possible to use the economic productivity of the whole resource base as the measure, rather than physical stocks of materials.[60] This would allow trading-off between technological progress and resource depletion. Unfortunately, much of the current evidence indicates that this trade-off does not often occur in practice (see p 97). It should also be noted that the trade-off between technology and resources may also be limited by the laws of thermodynamics (see p 100).

In conclusion, therefore, it might not make sense to preserve the natural capital stock completely intact if artificial capital were being steadily accumulated and could completely and satisfactorily substitute for natural capital, nor would it necessarily make sense to preserve the effective resource base (expressed as the sum total of natural and artificial capital, on some common numeraire) if technological developments were steadily increasing the economic value that could be produced from the effective resource base.

However, there is considerable uncertainty about all of these points. The prudent (and, given the current global environmental crisis, probably the most rational) strategy would be to ensure that any changes in the natural capital stock aggregate position were zero or positive. This would ensure that any improvements in the efficiency with which natural capital can be converted into artificial capital would represent clear gains. The alternative is to rely on the hope that such improvements will occur in the quantity and in the timescale necessary to compensate for losses elsewhere.

6

Integrating Economic and Environmental Factors

Any realistic set of policies for sustainability and sustainable development must make reference to all the relevant economic, environmental and other dimensions. It is likely that any real transition to a more sustainable way of life will require that economic and environmental considerations are integrated in practice as well as in principle.

This chapter looks in more detail at the various economic instruments that could be used to regulate patterns of resource demand and waste output.

Supply and demand

There are various ways in which economic and environmental considerations could be integrated. Conventional neoclassical economics can accommodate many of the immediate priorities, such as developing ways of valuing the environment. This valuation is necessary because, in the absence of other regulation, a resource that is undervalued, or even completely unvalued, will always tend to be used to excess. To understand why, it is necessary to review briefly the principles of supply and demand.

In general, the quantity of a good that people will buy depends to a significant degree on the price. The higher the price, the smaller the quantity that people will buy. This is partly because they will consume less (where possible), partly because they will find substitutes (where possible), and partly because some people will become unable to buy. This relationship between price and quantity is called the demand schedule or demand curve. There are some exceptions to these general principles. Where goods are to be given as gifts, for example, there can be social reasons why there must be a certain minimum price (in order to demonstrate an appropriate level of giving – there are some non-alcoholic drinks, for example, that are priced at a level that gives them equivalent gift value to a bottle of wine).

The amount of a commodity that manufacturers or farmers will produce will depend, similarly, on the price. As the price for a particular commodity goes up, more resources will be switched into manufacturing or growing that commodity. This relationship between price and quantity is called the supply schedule or supply curve. Supply and demand will tend to come to an equilibrium. If more is supplied than is needed, the price will be forced down. If less is supplied than is needed, the price will be forced up. The equilibrium price is the stable price at the point at which the demand and supply curves intersect, which is the point at which these forces are in balance.

If inputs of natural capital have low or zero values, therefore, more of them will be used than otherwise. This suggests that the price mechanism offers various ways in which natural capital demand could be reduced.

It is important to note that a regulatory approach would also require some kind of environmental valuation, as regulatory approaches require a planning framework, and priorities (which require values) are determined within such a planning framework.

There are a number of problems with the valuation of natural capital, however, some of which are reviewed in this section. Some of these problems may limit the extent to which environmental valuation can be used. Many of these problems arise because any values assigned to some element of natural capital will always be partly arbitrary, and so the setting of the value can never be completely uncontentious. It is true that there are some approaches (such as the tradable resource use and pollution permits reviewed in p 118) which avoid the need to determine precise values at the outset. However, this solution does not apply to all of the valuation problems.

The pricing of natural capital

In general, things become more expensive and harder to find, extract or process as they become scarce. So the scarcity of a good is reflected in the price, provided that demand remains at a sufficient level, with a premium obtaining for scarce goods. Anticipated future scarcity may also be reflected in today's price, as people will sometimes buy and stockpile against anticipated future high prices, and thus increase demand. In general, however, prices reflect current scarcity.

Economic scarcity, however, will not necessarily reflect every important aspect of the real state of the natural capital stock. Information concerning many elements of natural capital is sparse. Much of the necessary data has only been collected recently. It might be possible, for example, in the absence of adequate information, to use some form of renewable natural capital to the point where the stock collapses and its capacity to recover is destroyed. In other words, it could be possible for some elements of natural capital to be critical without necessarily being obviously scarce at the time of the transaction. The existence of a large number of individual members of a species, for example, might not be a sufficient guarantee of the long-term survival of that species if the genetic diversity within the group has been drastically reduced. In this case, the real resource being depleted would be the genetic diversity rather than the number of individuals.

Given the uncertain knowledge (see p 31) of possible ecological thresholds, it might also be possible to approach a threshold of system tolerance while the element of critical natural capital being depleted was still apparently abundant. This may prove to be true, for example, in some of the instances discussed in Chapter 4. Until that point, however, the real fragility of the resource might not be apparent. In other words, what would really be becoming scarce would be the capacity of the stock to recover, although that might not be reflected at all in the price.

It might also be possible to approach an important threshold without knowing, especially where the consequences are indirect. The continued existence of a given number of tonnes of ozone in the ozone layer might be sufficient to guard against intolerable rates of skin cancer amongst humans, but might prove to be insufficient safeguard against ecological damage if some important marine organism then turned out to be especially sensitive to UV-B radiation.

Similarly, it is probable that some elements of natural capital of unknown potential value, such as species, have been completely exhausted, that is, are now extinct, without there ever having been any knowledge of their existence.

This means that that the market price of forms of natural capital may not even reflect the real scarcity of that natural capital.

However, the problem is not just about the current lack of information about natural capital. One could imagine a perfect market, with perfect information available to all. In such a market, prices could reflect the 'true' value of every good. In the context of natural capital, however, one also needs perfect information as to the future. If one knew, for example, the cost of a good throughout all time, then one could 'collapse' this information across time by, for example, applying an appropriate discount rate. This would allow the good to be traded at a price that truly reflected its value to humans. If, for example, it was known that at some future point human existence would depend on the remaining supplies of some element of natural capital, this very high future value could be recognised in today's price, through some appropriate process of market correction. However, the future cannot be known, so this is impossible.

For example, oil is a non-renewable resource. It is sold at a price sufficient to cover the costs of extraction and to generate a profit. Imagine that at some future time someone discovers that some very valuable new chemical can be extracted from the oil. This future higher value of the resource cannot be properly taken into account, as it is simply not known in today's market.

The existence of competition creates another practical problem. No market operator could incorporate, for example, environmental externalities, or an allowance for compensating future generations for the loss of the natural capital, without becoming uncompetitive. So no market operator could act until all market operators were prepared or obliged to do likewise.

This means that markets cannot be relied on to generate 'true' values for natural capital. Other aspects of this issue are discussed on p 103 and on p 171).

Economic valuation of the environment

Environmental quality is not a market commodity. Various non-market based techniques are therefore employed to measure the economic value of environmental quality, all of which are problematic.

One of the key problems in establishing the economic value of environmental quality is in defining appropriate and observable measures of individual changes in utility or welfare. There is an even more fundamental problem with defining welfare itself, which is discussed on p 176.

Goods and services only have economic value because of their ability to provide consumers with satisfaction or utility. Utility cannot be literally measured, so it is necessary to find some other index of change in utility or welfare. There is an important problem as to whether a single index or multiple indices would be most appropriate. This is reviewed in Chapter 13.

Monetary measures are often used, because they are convenient, although it is clearly important to select the appropriate monetary measures. One monetary measure of net utility change is termed consumer surplus.

When people buy goods or services, they generally do so because they believe that their purchase will leave them in some way better off. That is, after deducting the actual purchase price, there will be some residual or net value. In general, the gross monetary value of a good or service can be thought of as comprising financial value and consumer surplus. Financial value is the actual amount spent on the good or service, and consumer surplus represents the residual difference between the gross value to the consumer and the financial value. Under some circumstances there might be no consumer surplus, so that the gross monetary value equalled the financial value, but the consumer surplus should not, in theory, be less than zero or the good or service would have negative value.

In practice, negative values may be possible. They could be incurred if people were to buy goods (presumably in error or under the influence of advertising) that were actually damaging or dangerous, that is, if they were mistaken in their estimate of the benefit. Moreover, people are generally unaware of the full opportunity cost of any purchasing decision or, even more significantly, pattern of purchasing decisions. Even if people were aware of all the trade-offs implicit in purchasing decisions, these are so many that it would be impossible to deal with them in a completely rational manner. It is also possible that real negative consumer surplus values are currently being concealed by the externalisation of some costs, such as environmental costs.

The financial value itself represents sacrifices or disutilities, so this disutility must be subtracted from the total utility to find the net change in utility derived from that good or service.

There are two main approaches to measuring consumer surplus, depending partly on whether financial value or utility is used as the dependent variable. Real data is more readily available with the financial value approach, but it is limited by the fact that it is necessary to assume that the price someone is willing to pay is a satisfactory measure of utility. The changes in utility approach is more satisfactory in theoretical terms, but creates a number of practical difficulties in measuring values.

The two approaches will, in general, provide similar results when the financial sums involved are relatively small in relation to incomes. When the sums involved become large in relation to incomes then the financial value approach will generate increasingly inaccurate results. This is because a given sum has different value for wealthy people or nations as opposed to poor people or nations. The sum may represent a modest reduction in the standard of living for the former, and an impossible choice for the latter. If poor people are obliged to choose between their survival and the protection of their environment, they generally opt to survive. Demand for environmental quality tends to rise steeply with rising income. This indicates that absolute levels of income are important. The changes in utility approach can allow for this factor.[61]

In practice, financial values are often interpreted as if they measured the consumer's welfare or utility. The value of a net change in welfare can be assessed, for example, by establishing the amount that the person would have to be compensated to make the change acceptable, or by establishing the amount that the person is willing to pay to bring about the change.

Given that environmental quality is not a market commodity, various methods are used to estimate values for environmental quality and therefore for changes in environmental quality. One is to estimate surpluses from actual market data, that is, from those costed behaviours that may reflect some aspect of the perceived utility of an environmental resource. For example, travel cost data is used to see how much people are prepared to pay or otherwise sacrifice in order to enjoy some particular area. The hedonic price method uses multiple regression analysis to extract the effect on some market price of some aspect of the environment. An attractive setting, for example, tends to increase house prices. The results of multiple regression analysis are generally sensitive to slight variations in the initial conditions, which means that results require interpretation and judgement. This in turn means that this technique must be used with especial care, and that it is difficult to use this technique for any cases other than relatively simple and uncontentious ones.

However, there are likely to be some aspects of environmental quality where this approach cannot be used, as there is no relevant market data. The approach that has been generally adopted here is to set up experimental or artificial markets. People are asked to imagine that the aspect of the environment in question were traded, and then asked how much they would be prepared to pay or what they would be prepared to sacrifice. Some studies have used actual payments in these artificial markets, others have used contingent or imaginary payments.[62]

One approach, for example, would be to set up an imaginary public referendum in which people are asked to indicate how much they would be prepared to pay to improve some aspect of environmental quality.

There are a number of problems with the experimental market

approach. Comparison across a number of actual studies done in this area indicates that there are a number of extraneous factors that can significantly affect the results.[63] The precise definition and description of the environmental variables, the general level of information available to participants, the initial selection of the measure of welfare to be used, the techniques selected for the subsequent analysis of the data, the selection of the participants in the experiment (especially with regard to their income and educational attainment levels), the non-respondents problem (people who choose not to cooperate in such trials may be doing so for important reasons, which it would be important to understand, but which cannot be extrapolated from absent data) and a number of other methodological and procedural factors can all have significant effects on the results obtained.

The range of responses obtained is often vast, and may range from zero (unwilling to pay anything) to infinite (willing to pay anything) in a single study.[64] It is of limited value to derive an average response from such a range. It may even be entirely misleading, as a single average might conceal a multi-modal distribution. If the population, for example, contained two groups, one of which placed a high value on the environment and the other one of which did not, an average might well fall between the groups, on a value supported by none of the respondents.

The only results obtained by this approach that can be agreed with any certainty, therefore, are:

❑ Most people are probably prepared to pay something or sacrifice something for the environment.
❑ The amounts involved may be quite large in some cases, but the range of responses is typically wide. At least the responses obtained are not negative values (people are not generally prepared to pay to have their environmental quality reduced), but, apart from that, not much is as yet established.

Environmental property rights

The previous section introduced the idea of economic valuation of the environment. If the values derived from such procedures are then to be used as the basis for some sort of compensation process, so that people could be given some form of recompense for a deterioration in the quality of their environment, there would also have to be environmental property rights (see p 111). People are not, in general, compensated for the loss of something that they do not own or have no rights over. If people are to be compensated for a deterioration in environmental quality, it must be assumed that they possess some right to a certain standard of environmental quality. Similarly, the existence of a market in any kind of environmental capital, either in resources or in pollution absorption capacity, would depend, as do all markets, on the introduction or recognition of the relevant property rights.

There are some precedents. Some planning law, for example, has had the effect of removing environmental property rights from landowners and developers and re-assigning them to society as a whole. Regulations that govern emissions have a similar effect.

The extension of the concept of environmental property rights would be a complex procedure. It would be necessary to establish, in quite diverse contexts, who should hold such property rights.

Where there is little definition or enforcement of environmental property rights, they are usually effectively owned by the polluters. If these rights were to be transferred to those affected by pollution, then polluters could be obliged to pay to use the environment as a waste disposal option. This would have the effect of making other waste disposal options more economically attractive, some of which might also be more environmentally sound.

There would clearly be some political difficulty in developing this idea internationally, as the necessary structures to define and enforce these environmental property rights do not currently exist. Without such structures, it would be difficult to prevent abuses (see p 172).

The extension of the idea to truly planetary resources, such as the ozone layer, would present similar complexities. It might be necessary, as a precondition, to develop appropriate global structures for the cooperative management of the Earth (see p 175).

Environmental quality and sustainability

The complexity of the issues involved in the question of sustainability makes it important to consider how and what axes are chosen to assess movement towards or away from sustainability. It is especially important to note that the assessment of sustainability will not necessarily require the same measurements as those needed to assess, for example, perceived environmental quality. There may well be some overlaps, but the questions are not identical. In some cases, overlaps will mean that the same measurements may be taken, but that the interpretation will be different.

This means, for example, that when taking potential habitat or species losses into account it is important to define and apply an appropriate set of values. There is, as yet, no evidence that human preferences expressed, for example, through the market place relate to the relative criticality (as opposed to the simple economic scarcity) of forms of natural capital. In principle, this could be due to the lack of information on the subject, which could be corrected. However, there are some indications that certain human value systems are actually quite unhelpful in this regard. For example, people tend to place a higher value on what they have than on items of equivalent value that they do not currently possess.[65] Thus people generally have some tendency to resist change, no matter how desirable that change might be in some larger context. Similarly, people tend to value their environmental status quo rather than a healthy ecology (people have objected to the removal of factory chimneys and slag heaps, and to the reafforestation of denuded hillsides). People also tend to be more aware of, and so give higher priority to, environmental change that happens quickly rather than slowly, although long-term environmental change can be more serious. Finally, there is an understandable human tendency to assign higher value to species that live at or near the top of the food chain, partly because that is where we ourselves are and partly because these will always be fewer in number than species at or near the bottom of the food chain (see p 84). For example, people tend to get more exercised about elephants than about blue-green algae, and will donate to save pandas rather than beetles. This anthropomorphic set of values will not necessarily reflect the contribution made by a given species to the maintenance of the global ecology. Blue-green bacteria play a more significant role in this regard than elephants, or, for that matter, humans. Any valuation procedure based on anthropomorphic values will provide information about current fashions in human likes and dislikes, but may not necessarily provide useful information about sustainability (see also p 171).

7

Economic Policy Instruments

If a country's labour and capital stock were being used as intensively as possible, then it would be impossible to put labour and capital to any new or additional activity without taking them away from somewhere else. This is referred to as being at the production–possibility frontier.[66] If the labour and capital in the UK economy were fully employed in this way, then it would be impossible to divert resources from steel fabrication into electronics, for example, without some reduction in the output of steel. However, the increased output of electronic goods might be worth more than the lost output of steel. Resources could be progressively diverted until the point at which the additional output of electronics would be worth less than the lost output of steel. Resources are said to be allocated optimally at those points at which no further switching of resources could result in a net increase in value. This is often referred to as being at the margin. This is the point at which the switch of the last (or marginal) unit of resources results in a net loss of value or welfare.

In practice, of course, most economies are not fully employed in this way. This is referred to as being inside the production–possibility frontier. In such an economy, growth in output can occur by taking up the slack, that is, by putting unemployed labour or capital to work. This does not necessarily entail any loss in output from other sectors of the economy (though in practice it can, because of the unequal distribution of assets such as skills and knowledge).

The maximum possible total value of goods and services will be produced, with maximum efficiency, when a society is at its production–possibility frontier and resources are optimally allocated. It is generally true that this maximisation will not usually be achieved in a market economy. This is for several reasons:

❑ Society has wider concerns than the optimal allocation of resources. The way in which resources are allocated will affect other factors, such as the distribution of income and wealth. Society may choose a sub-optimal allocation of resources in order to achieve certain changes in the distribution of income and wealth.

❑ Competition is rarely perfect. Market operators usually compete on unequal terms for the available resources. These outcomes depend, at least in part, on existing inequalities, rather than on what would be an optimal outcome.

❑ Markets depend on people, and people depend on information. Information, however, is never perfect, and people make both errors and irrational decisions.

❑ There are some types of goods, called public goods, to which access cannot be controlled. For example, once a lighthouse is installed, any mariner may use it for a bearing. These goods cannot, therefore, be supplied at the socially optimal level by the market. There are also types of goods, called merit goods, which are included in the group of public goods on ethical grounds. For example, although it would be possible to ensure that only those who could afford medical treatment received medical treatment, many people believe that this would be unethical.

❑ Some (possibly most) types of production externalise some of their costs. Optimum output cannot, therefore, result until some market correction re-internalises these costs. This is a particularly important factor in environmental economics.

The production of goods and services operates in a similar way to that described earlier: goods and services will be produced up to the margin, the point at which the value of any further production is less than the total cost of obtaining that additional production. The individual or the firm will have an incentive to produce goods or services until the point at which the price they receive for the marginal unit is less than the cost to them of producing that marginal unit (see the discussion of the principles of supply and demand on p 106).

One of the central problems in environmental economics is that these private costs to the individual or to the firm will not equal social costs. This is because some of the costs of production are not being borne by the producers. If natural resources are undervalued or unvalued, this will reduce the cost to the producer who uses these resources as raw material. This will then encourage the producer to use more than they otherwise would. Similarly, if producers are allowed to use low-cost waste disposal options, which cause pollution, this reduces the cost to them. This will then encourage the producer to produce more waste and pollution than they otherwise would.

Society, however, suffers a loss of welfare in both cases. Additional natural resources will be appropriated, and so not available for other purposes, and additional pollution will be generated, which will cause some loss of amenity or perhaps more serious damage.

It is important to note that if a firm continues to produce goods to the point where its own costs are covered at the margin, the total cost (which includes the cost borne by the firm plus the cost borne by society) must exceed the value of the marginal output unless there genuinely are zero costs to society.

This difference between private and social costs will invariably lead to a misallocation of resources, and therefore a sub-optimal outcome for society.[67] The firm will have an incentive to produce more of the goods and services than would otherwise be produced, because their output is being subsidised by society. The resources required to produce that output are not, therefore, being optimally deployed. If the subsidy were removed, output of the good or service concerned would drop, and some of the resources tied up in that production would therefore become available for other production which would then be of higher value.

Environmental taxes, charges, and subsidies

One possible solution is to uses taxes or subsidies to bridge the gap between private and social costs, and thereby correct the misallocation of resources.

Firms would then increase their output to the point at which their internal costs plus the tax would equal the price of the product at the margin. At this point the total social costs would also be equal, at the margin, to the value of the output.

This is why environmental taxes, almost uniquely, would actually improve the economic allocation of resources. Most subsidies and taxes distort the allocation of resources. For example, income tax relief on mortgages and agricultural commodity price support are resource misallo-cation measures. These subsidies ensure that more of the output (owner occupiers and food, respectively) will be produced than is economically optimal, by absorbing resources that would be more productively deployed elsewhere.

Environmental taxes on inputs or on certain outputs (such as pollution) could be used to replace other forms of taxation. Thus the effect of introducing environmental taxes could be revenue-neutral, while the economy itself would function more efficiently as resources would be better allocated. In fact, if other taxes were not reduced or withdrawn as environmental taxes were introduced then the economy would deflate, unless public expenditure were increased at the same time to return the additional revenue to circulation. It has been suggested that the revenue from environmental taxes be hypothecated, that is, used to fund additional public expenditure on environmental clean-up operations, but in economic terms there would be no necessary connection between the source of the revenue and the expenditure. A degree of hypothecation might, however, be desirable in political and social terms.

There are other reasons why the introduction of environmental taxes, charges, or subsidies would be preferable to the introduction of direct controls. In principle, both direct controls and environmental taxes could be used to regulate pollution and resource consumption. Both methods would require that flows of pollution and resources be monitored and measured. However, once these measurements had been made, taxes would probably be the more cost-effective means of control. This is because the implementation of regulations involves some cost, while taxes are usually supposed to raise more money than they cost to collect.[68, 69]

Such taxes could be related to, for example, the volume and strength of the pollution. This need not necessarily require very precise measurements and complex formulae for comparing across types of pollution, as broad bands would be adequate in many cases.

In addition, the regulatory method does not create an economic incentive to achieve higher standards than are required by the current regulations, although some firms might try to achieve better than current standards in the face of consumer pressure, or where the firm anticipates further raising of regulatory standards or is prepared to subordinate short-term profits to a long-term gain of market share. With environmental taxes, however, there is always an incentive to reduce costs by further reducing the use of those inputs or the emission of those outputs for which a charge is incurred. For example, energy and resource taxes would continuously encourage all operators to improve energy productivity and resource use efficiency, and continuously discourage them to permit inefficient practices that waste energy or materials.

Of course, it is possible that producers would attempt to evade the tax or pass on some or all of it to consumers by raising prices. The extent to which this would happen would depend partly on the level of tax, the cost of tax avoidance, and the degree of competition in the market. However, any significant increase in price to the consumer would have the effect of driving down demand, which is undesirable from the point of view of the producer. The need to control costs and so maintain profit margins would therefore provide a continuous incentive for improvement.

Another way of looking at this is that prices are generally related to the benefits they confer on the purchasers (see p 109). These prices can be positive or negative. With positive prices, the more one sells the more one gets. With negative prices, the more one sells the more one pays. Taxes are a form of negative price. Many forms of resource use and pollution do not currently carry a negative price. There is no relation, therefore, between the price and the associated damage, and therefore no disincentive to this kind of activity. Similarly, there is no price incentive to develop, for example, technologies that would reduce pollution. This is a very important point in the argument, reviewed on p 93, about substitutability and technology.

There is an exception to the general suggestion that environmental taxes and charges are preferable to regulation. That is where some element of natural capital is so critical that no use is prudent, or where some ecological balance is so finely balanced or near collapse that no damaging emissions can be allowed at all. In a sense this could still be incorporated in a single system of taxes and charges, by allowing that in some circumstances the charges would rise towards infinity and thereby make the action concerned impossible, but in these situations direct regulation is likely to be simpler and more enforceable.

There would probably be a number of transitional problems with the introduction of such environmental taxes. One of the most important would be the redistributional effects. Some firms, sectors, or regions would gain, but others would lose. Of course, it would be possible to introduce other policy instruments at the same time. In general, it is often more efficient to have separate policy instruments, each with a specific goal. In this instance it would probably be preferable to optimise resource allocation in the economy with pollution and resource-use taxes, then to deal separately (but as part of the same overall package of measures) with any consequently necessary measures to redistribute income. In this instance, associated compensatory policy measures might include the following:

❑ The losers could be compensated directly. This, however, is not always easy to implement or politically popular (apart from with those people being compensated).
❑ The taxes could be relaxed for a particular sector or region. This tends not to be politically popular with other sectors and regions, and tends to weaken the benefit of the change.
❑ It would be possible to phase in such taxes, and to allow a transitional period. This solution is frequently used when new measures are introduced, in order to allow people time to adjust.

Of course, it is important to note that any redistribution problems would arise during the transition from the existing distribution. It is quite possible that the existing distribution is itself is both inefficient and unjust. Pollution, for example, represents a transfer of welfare from the victims to the perpetrators.

There is a related set of arguments as to whether to use taxes or subsidies. The government could decide either to tax firms that perpetrate undesirable practices or to subsidise firms which avoid any undesirable practices. The equivalent level of tax and subsidy will represent exactly the same opportunity cost to the firm concerned. To be charged a sum represents the same opportunity cost as a sum that is foregone. These outcomes are not generally perceived as being equal, however, so cannot be expected to have the same effect in terms of motivating the desired response. Furthermore, taxes and subsidies have different distributional effects. A pollution tax, for example, would have the effect of improving the welfare of the taxpayer and decreasing the welfare of manufacturers, consumers or both. The exact equivalent subsidy would increase the welfare of manufacturers and consumers and decrease that of taxpayers.

Of course, subsidies need not be direct. Compensating projects (see Chapter 9), where public expenditure on environmentally enhancing projects is used to compensate for environmentally damaging private projects, represents a form of indirect or general subsidies to the private sector. (It would also be possible, of course, to fund such compensatory projects directly with appropriate levies on the damaging activities).

Tradable pollution and resource use permits

Pollution diminishes environmental quality. It can therefore be thought of as using up part of something that has value. Insofar as pollution is necessary to production, environmental quality can be regarded as a scarce resource, and thus a form of capital. Since capital is a factor of production, this suggests that environmental quality should also be regarded as a factor of production. The price mechanism is used in market economies to allocate labour, raw materials, and the other factors of production. By simple extension, the price mechanism could be used, at least in some situations, to allocate pollution.[70]

The idea of tradable pollution permits is based on this argument. First, there would have to be some decision as to what the maximum allowable level of pollution should be. This maximum aggregate emission could be determined on the basis of the best scientific evidence available, or could be set at any other level that society wished.

Pollution permits whose value summed to that amount could then be issued and sold. Any firm that generated pollution would have to have the appropriate number of permits, which they would have to buy in each year. This would ensure that those polluters who could reduce their emissions cheaply would do so, while those polluters who could not reduce their costs cheaply, because they had old inefficient plant for example, would have to bid for permits instead. How this process might work internationally is reviewed on p 172.

One strong advantage of this approach is that it avoids the need to try to cost the effects of pollution. The price of the permits would simply be determined by the market. Of course, if this is to be used as a revenue-raising measure, the initial price must be determined by the government, but the permits could then be traded in a secondary market at purely market-set prices.

This is important, as it is difficult to try to cost the relative values of the goods produced and the pollution caused by their production, partly because of the complex methodology required and partly because of the value judgements involved. Existing social arrangements and structures, for example, affect the relative price of goods. Where there are private monopolies, prices tend to be higher than where there is competition. This means that if the output from a monopoly was threatened by pollution this lost output would appear to have a higher value than if the lost output (which could be otherwise identical) was being produced by competing firms.

This principle applies in several contexts. Where waste disposal is a monopoly, the cost of waste disposal is likely to be higher than it would be if it were open to competition. If proper waste controls were introduced in such a situation, and if these controls actually cost some 'output' (disposed waste), then the value of the lost output and hence the apparent cost of the introduction of such controls would be higher than in a situation where there were firms competing for the business.

If tradable permits were introduced, then the costs at the margin would be equal for all firms. This is because pollution abatement would switch from firms where it would be costly to those where it would be cheap. This would be likely to have large redistributional effects, with the benefits and disbenefits relocating in society (relative to the current situation). Again, specific measures could be introduced to provide compensating effects, if that was felt to be necessary or desirable.

Exactly the same approach could be used to control renewable resource use. If there were some maximum annual harvest that some natural resource could provide, then licences to that amount could be issued, and the market would determine the price.

There are, however, much more complex problems with non-renewable resources. It may not be possible to use a tradable resource-use permit approach. This is essentially because it is much more difficult to estimate possible future values of non-renewable resources. For example, non-renewable resources could rise in value over time as increasingly costly operations were required to extract poorer or less accessible reserves, until reaching the point at which reserves could not be economically extracted (see p 91 and p 125). It is also possible that some higher-value uses of non-renewable resources remain to be discovered, which could not be anticipated by the market today.

It is clear that a tradable resource-use or pollution permits approach could only be applied to current pollution and resource demand, and that other measures would be needed to deal with the inheritance of exhausted resources, degradation and pollution left by previous generations.

There are two main approaches to the problem of inherited pollution:

❑ Public works, to deal with the problem collectively.
❑ Some mechanism to assign responsibility for the problem, perhaps quite arbitrarily. The US, for example, has established the principle of site contamination liability with the 1980 Superfund Act, under which site owners are liable for the costs of decontaminating sites even if they were not responsible for the original contamination. This has had some far-reaching consequences. The Environmental Protection Agency in the US has estimated that there are currently some 27,000 hazardous waste sites in the country, which will cost about $675 billion to decontaminate. It is likely that in some of these cases the clean-up cost will exceed the value of the site, and possibly of the entire operation or even the firm. This would mean that some loans, currently secured against property, may prove to be secured against net liabilities. Similarly, the European Union has considered legislation to oblige a firm's bankers to assume responsibility in those cases where a firm defaults or becomes bankrupt and leaves unmet environmental costs. This may oblige banks to insist on environmental audits of some of their clients.

The value and cost of pollution reduction

The argument that the environment can be regarded as a scarce resource and pollution as a factor of production that uses up some of this resource has been extended by some to support a supplementary argument. This is that pollution should only be reduced to the point at which the costs of any further reduction would be greater than the monetised value of the benefit derived from that reduction.

There are indeed situations where the costs of preventative action rise steeply at the margin. To reduce the heat loss from a typical UK house by 30 per cent is relatively simple and cheap. To reduce the heat loss by 90 per cent would be relatively complex and expensive. This relationship between expenditure and reduced heat loss is not linear. To move from a 20 per cent to a 30 per cent reduction is often cheaper and easier than to move from a 80 per cent to a 90 per cent reduction. Similar relationships obtain in pollution control, where relatively large reductions can often be achieved with good management, while relatively small reductions beyond that stage, to near-zero emissions for example, sometimes involve a major re-organisation and expensive new technology.

In such cases, the appropriate trade-off between environmental quality and the economic cost of reducing pollution may be fairly obvious. However, such decisions are best made in a strategic context. If the firm anticipates that environmental standards will rise, it may well make sense to commit the necessary expenditure and upgrade to the higher standard from the outset, rather than being faced with a demand at some future point to upgrade to the higher standard and to write off the cost of all the more modest incremental improvements made earlier.

It is also important to apply an appropriate set of values when assessing the value of the benefits derived from reductions, which is what determines the optimal level of expenditure. Human values do not necessarily coincide with ecological values (see p 112). People sometimes place a higher value on visible change (such as an obvious visual intrusion), for example, than non-visible change (such as chemical changes in the soil). Thus some have argued against the development of wind turbines on the grounds of their visual intrusion, and thereby implicitly accepted the continued operation of fossil-fuel burning power stations with the attendant deposition of oxides of sulphur and the acidification of soils and waters.

Similarly, the ecological value of any reduction in pollution will depend partly on the sensitivity and current condition of the environment. A 10 per cent reduction in output into an environment near a threshold of ecological collapse may be much more significant, in terms of averting this damage, than an identical 10 per cent reduction in output into a robust and healthy environment.

8

Discounting and Investment

Following on from the discussion of economic instruments in Chapter 6, and the use of environmental taxes, charges and subsidies, and tradable pollution and resource-use permits in Chapter 7, these next two chapters look in more detail at valuation issues. This chapter considers investment and the discounting of future values, while Chapter 9 reviews the problems with identification and integration of costs and benefits, which may in practice be distributed across different projects, countries, or generations.

Capital, investment and net productivity

Capital can have net productivity. This means that there can be a real return on investment, over and above the cost of the investment. Sensible investments in machinery, transport infrastructure, and training, for example, can lead to real increases in output later on. Some investments have high net productivity, others low net productivity.

In general, there is more scope for 'big' projects with high net productivity when economies are going through particular development phases. This is why, for example, fortunes were made in steel and railroads when the US was industrialising around the turn of the century. Later on, when most of the big projects are completed, there tend to be fewer opportunities for such productive and profitable investment, and returns on investment diminish. This is why, if capital is mobile, funds then tend to go overseas into investments in more rapidly developing economies.

All societies have a limited stock of capital and capital goods. There therefore has to be some way of prioritising investment projects. One common strategy is to prioritise those projects that have the highest net productivity. One of the simplest ways to do this is with interest rates. People do not generally borrow money to invest unless there is a reasonable prospect that the return will be greater than the interest rate. High interest rates therefore filter out all but those projects that offer a potential return that is yet higher. Of course, lenders must lend in order to make their profits, so interest rates and net productivity rates tend to co-vary, unless either or both is being managed as part of some government policy (such as the policies in the UK from 1987 to 1992 of having relatively high interest rates with simultaneous controls on public sector investment).

The interest rate can be thought of, then, as the market price of investment funds. Its level is determined by the interaction of supply and demand, and by controls exerted by central governments and banks.

If net productivity is generally high, returns on investment will be high, which will tend to keep interest rates high. As a society completes a phase of development, net productivity starts to fall, and interest rates, other things being equal, should come down too. If net productivity is low, there is relatively little financial incentive to invest, although there would still be some investment even at zero interest rates. It would still be financially worthwhile to spend money to insulate one's house, for example, as long as the savings in fuel exceeded the cost of installation.

Reasons for discounting

Investment options are usually compared in terms of the revenues that each investment is expected to provide. The general principle of discounting is that future revenues are discounted because they are assumed to be less valuable than current revenues. This, of course, tends to favour investments that offer greater short-term returns and militate against long-term investment.

One of the most important reasons for discounting is the existence of net productivity and interest rates. Capital invested today, at interest, will earn a profit. If the capital does not become available until tomorrow, then today's potential profit is lost. This is why high interest rates generally co-vary with high discount rates.

Discounting is also used for various other reasons. One factor is time preference. Under prevailing circumstances, people generally prefer to have returns on investment sooner rather than later. This is for a number of reasons:

❑ People are impatient.
❑ A given sum of money generally makes more of a difference to someone who is poor than someone who is rich. A trivial sum to a billionaire might seem like great wealth to someone on welfare. If an individual or a society is becoming or expects to become wealthier, therefore, it is reasonable to suppose that a given sum will make less of a difference to them in the future than it would today. Thus people prefer to have their returns sooner rather than later, as they anticipate that the later income will be worth less to them. This is termed the diminishing marginal utility of income.
❑ There is a risk to the investor that he or she might die before a long-term investment matures.

Clearly, there are a number of factors which make time preference entirely rational behaviour, and which tend to shorten the acceptable duration of an investment period. For example, conditions of extreme poverty oblige the sacrifice of long-term interests in order to secure short-term survival. Similarly, time preference tends to become more acute in times of social and economic uncertainty, as one cannot assume that any long-term investment will be safe.

The discounting debate

As long as society is growing wealthier, as long as there is net productivity, as long as there are interest rates, as long as people are impatient, and as long as people value financial returns, they will discount. The fact that people do so is not something that could or need be controlled. Discounting is simply a means of prioritising investments and allocating resources.

There is a debate as to whether discounting is compatible with sustainability. Some of the issues are reviewed in the remainder of this chapter. In general, discounting appears to make it more attractive to defer costs, which could be environmental costs, to future generations. Furthermore, should it prove necessary to finance a long-term reconstruction of the industrial base of society, discounting might make it more difficult to allocate the necessary capital.

There are unresolved arguments as to whether discounting itself really poses new problems, or whether discounting, when applied to monetary values that are supposed to represent environmental factors, simply compounds some of the problems associated with the monetisation of environmental factors (which are dealt with in various chapters in this book). It is clear that, at the very least, these issues become much more complicated when discounting methods are used.

The validity of the assumptions

As suggested earlier, time preference in circumstances of poverty and uncertainty is quite understandable. However, there are important questions about the role of time preference in a relatively wealthy and stable society.

❑ Impatience is not necessarily the best basis for prudent asset management. Long-term investment may (and often does) lead to greater returns over the total life of the project.
❑ The diminishing marginal utility of income depends on the anticipation that society will continue to become increasingly wealthy. If that proves to be incorrect, then income will not diminish in value in this way. It is impossible to be certain that any society will invariably become more wealthy. There have been a number of instances in

DISCOUNTING AND INVESTMENT | 125

which societies have become poorer. In a world with a growing population, in which environmental constraints and resource extraction costs may become more significant, and in which there may be growing pressure to reallocate access to resources, there must be a possibility that at least some societies or some sections of society will become poorer (see Chapter 4). Under such conditions income would not diminish in value, and there would be some investments which would become more rather than less relatively expensive (and therefore less rather than more possible) over time.

❑ The mortality factor, however, remains. Attempting to minimise the risk of losing revenues that might accrue after one's death does constitute rational behaviour, if one's goal is to maximise personal revenues. This kind of behaviour is particularly understandable in a society that operates on a principle of intergenerational inequity. In a society that has had resources appropriated by past generations, and seeks to obtain the maximum share of resources in its turn (and thereby appropriating some part from future generations), it is quite defensible to seek to minimise one's 'losses' to future generations. In a society that applied sustainability constraints over a number of generations, for example, it would be necessary for each generation to accept the 'cost' of leaving resources to future generations. Each generation would 'gain' in return the resources that the past generation would have left to their future generation. Unfortunately, this situation is exceptionally open to one-sided cheating, as the future generation have no recourse, and there would have to be some cultural code of behaviour with appropriate reinforcement and sanctions to serve as a substitute. However, in principle, the risk of the mortality argument could be mitigated in such a society. It would be possible to accept that future generations would derive the main benefit from one's investment, because one would be the main beneficiary of the investments of a prior generation. Such cultural reinforcement would enhance the perceived 'bequest value' component of investment returns. Of course, this is little consolation to the first generation that attempts to break out of the cycle of repeated inequities, as some of the resources of that society have already been appropriated.

Discounting, society and the environment

Discounting is a rational way of deciding how to allocate costs and benefits over one's life. However, the application of discounting principles is not equally appropriate in all contexts, including some of the contexts in which discounting is currently used. In some cases the inappropriate application of discounting can give rise to inefficient, dysfunctional, or socially and environmentally undesirable consequences. There are a number of issues around this point, which are briefly reviewed below:

❑ In general, discounting makes sense for an individual or an individual business, as people at that level are legitimately concerned with their own returns. There is a widely accepted principle that society as a whole can and should take a more long-term view, partly because society generally outlives most individuals. However, while many would accept that there should be a 'social discount rate', there is as yet no consensus as to whether, in the context of sustainability, this rate should simply be lower than the personal discount rate or whether there may be cases where it is more appropriate to disregard the distribution of benefits over time and to apply a zero discount rate. In fact, if resource costs are rising, or if a great deal of environmental remedial work must be undertaken, it might even be more appropriate in some instances to apply a negative rather than a positive discount rate, or, more conservatively, to assume that future generations will at least derive no less benefit than current generations from natural resources and to apply a zero discount rate.

❑ A number of serious issues arise when environmental factors are assigned a monetary value, these values are then discounted in various cost–benefit analyses, and the decisions that result are then translated back into real-world consequences. For example, one procedure for comparing the value of a resource to current and future generations is to estimate a monetary value for future utilities or disutilities and then to apply a discount rate to express the present capitalised value. However, it is possible that the process of monetisation might fail to fully capture the entire range of values of the environmental input. This could, in itself, be highly significant. However, the fault may become more significant if discounting methods are used. In particular, if any one or more of the values not captured in the process of monetisation actually increases over time, the outcome of the cost–benefit analysis would become even more at variance with reality, and with the most efficient or otherwise desirable outcome.

❑ Discount rates affect the distribution of both costs and benefits over time. In general, benefits are preferred in the short-term (positive discount rates mean that long-term benefits are taken to be less valuable), while costs are preferred in the long-term (because they are taken to be less expensive). It is possible that this process could generate some intergenerational inequities. With a nuclear power station, for example, one generation might obtain the electricity while bequeathing the nuclear waste disposal problem to the next. If the electricity consumed has allowed the creation of wealth in some form which the next generation also inherit, they may feel that they have been adequately compensated. This depends, however, on whether the compensatory wealth is actually created at a sufficient rate. If the demand for the electricity is miscalculated, the electricity is consumed inefficiently, the profits invested unwisely, and the benefits thereby dissipated, the net productivity of the project may be low, so that the future generation actually receive little compensation.

❑ High discount rates erode the value of long-term investments. They therefore tend to encourage short-term investment and place a premium on current consumption. If it were necessary to encourage a shift towards a more long-term attitude to both consumption and investment, it might be necessary to operate lower interest rates, or a reduced rate of future discounting. However, if it became necessary to encourage a diversion of capital from current consumption into saving and investment, it would be necessary to ensure in addition that the return on investment was above the personal rate of utility discount for lenders, or people would simply consume rather than invest their capital.

❑ It has been argued that resource depletion is justifiable if a 'sufficiently large' proportion of resources, having been transferred into artificial capital, is reinvested or held in the form of infrastructure instead of being consumed in the present. This invested capital is then supposed to be sufficient compensation for future generations (this, of course, depends on the assumption that artificial capital can be an effective substitute for natural capital). On this basis, a positive discount rate is used when comparing the utility of a resource to a future generation compared to the current generation, which has the effect of giving the current generation preference over future generations. Such positive interest rates then provide an incentive to work resources to exhaustion (provided that the rate of increase in their price in situ is less than the interest rate). There is therefore both a rationale and an incentive to exhaust resources today, leaving artificial capital as compensation for future generations.

There are a number of problems with this compensation argument. One is the existence of conversion inefficiencies. The second law of thermodynamics means that the conversion of natural capital to artificial capital will incur real losses of energy and material, although the utility may have been increased. For some purposes, this restriction can be avoided by utilising 'free' inputs such as solar energy (which is not actually thermodynamically free, but can be considered to be so for practical purposes). If the compensation is calculated in terms of utility, then these real losses may not appear to matter. However, under circumstances of increasing resource shortages and rising resource costs, such real losses might become more significant.

Another is that it cannot be assumed that future generations would wish to make the same use of the resource. For example, the utilisation of oil as a fuel will probably become economically non-viable before its utilisation as a chemical becomes economically non-viable. In this case there would be a trend from low-value uses to high-value uses. There are a number of factors that suggest that this will be a general trend, and that resource values will be higher rather than lower in future. Broadly, these are as follows.

– Technological developments may raise the utility of resources by making higher-value uses possible.

– Increasing resource costs would encourage the trend to these

    higher-value uses.
- Some resources may prove to be non-substitutable by technological or other means.
- Some resources may become increasingly scarce in real terms.
- Energy costs may become higher in real terms.

❑ Another variant of the discounting and compensation argument is the theory that a resource should be exploited as long as the opportunity cost of not exploiting that resource exceeds the sum of the current non-use values plus the estimated future non-use values. Thus a forest should be cut down, for example, as long as the value of the timber exceeds the sum of the present and future values of the forest for conservation of biodiversity, recreation, amenity and so on. This approach, however, leads to an anomaly. The opportunity cost of not developing a resource is the sum of the maximum values placed on non-exclusive uses or the maximum value of the highest value exclusive use. This means that every additional use of a resource adds to the opportunity cost of not exploiting a resource, thereby making it less likely that the resource will be conserved. For example, oil can be converted to plastic. If this plastic were in the form of some disposable item, and used only once, then the opportunity cost of not using the oil would be relatively low. Every time the plastic is recycled and re-used, the opportunity cost of not using the oil goes up. Recycling therefore increases the 'cost' of leaving oil in the ground. According to this model, therefore, improving the efficiency of resource use would increase the rate of resource exploitation.

❑ Another approach to the compensation problem is to estimate the amount of money that would have to be set aside from the proceeds of the liquidation of a resource to generate a permanent income stream that would be as great in the future as the portion of receipts that are to be consumed in the present.[71] For example, imagine a mineral deposit that could be worked out completely in a few years. The earnings from this, which are finite, have to be converted into an infinite series. Every year, some part of the revenue can be spent on consumption while the remainder (the capital element) is set aside and invested in order to create a perpetual stream of income that would be equivalent to the current level. This level is termed the 'true income' (which means equivalent real incomes over time), and will hold a certain ratio to total receipts. There are some problems with this argument too.
- The number of years to exhaustion is not a simple calculable fact, but partly a function of the way in which the energy cost of extraction and the relative value of the mineral vary in relation to each other. It is impossible to be certain how these factors will vary in future. It is therefore impossible to be completely certain as to how many years the resource will last, and how long the extraction profit income will continue to flow.
- This approach depends on the assumption that the price of non-renewable natural capital will remain constant in relation to the general price level. If there is a relative increase in resource prices,

then the amount to be set aside would need to be a proportion of the future price rather than the current price, which would then make it necessary to try to estimate what the future price might be.

- If the discount rate is 4 per cent, and the life expectancy of the resource is 35 years, then the set-aside rate is 25 per cent. If resource prices rise by a factor of four relative to general prices the set aside becomes 100 per cent. If the discount rate is 0 per cent then the set aside is also 100 per cent. This does not necessarily preclude any developments, but it does mean that under some circumstances it would be necessary to include 'compensating projects' (see Chapter 9), which would have to be added in to give a combined compensation of equivalent value to the resource.

❑ All artificial capital depreciates. If some future generation inherits an artificial capital asset that then requires a lot of maintenance, or has a relatively short remaining lifespan before decommissioning and disposal, this may not at that point represent an equitable exchange. It would even be possible for this inheritance to be a negative asset if the recurrent maintenance costs exceeded the use value, or if the pattern of demand had changed so that this form of artificial capital was no longer required. For example, the US has invested resources in the construction of a national network of roads. Much of this network is weathered, and the annual maintenance costs are starting to climb. If fuel costs now increase, and private car use becomes less attractive as the base transport policy, the legacy of a crumbling and economically unattractive motorway system may not appear to be as equitable an exchange as formerly.

❑ The use of discounting over generational boundaries is ethically dubious. While it is arguably legitimate to allow anyone to decide how they wish to apportion their costs and benefits over the course of their lifetime, it cannot be equally ethically legitimate for anyone to allocate the benefits to themselves and the costs to other persons or to future generations.

❑ Discounting is a sensible way of deciding how to allocate financial costs and benefits over relatively short periods of time. The standard discount period used in industry is rarely more than ten years long, with the most notable exception perhaps being forestry. However, discounting over much longer time periods can give rise to absurd results. With ecological impacts that may be measured over thousands of years, a difference of a small fraction of a percent in the discount rate used may determine whether the answer to a cost–benefit analysis is favourable or unfavourable. This is too open to possible bias to be uncontentious.

Discounting and natural capital depletion

The relationship between discount rates and the rate of natural capital depletion was discussed in the previous section. Changes in interest rates will also affect rates of natural capital depletion, and controls on rates of natural capital depletion will affect interest rates, depending on a number of factors such as technological development and the extent to which artificial capital can substitute for natural capital. Consider the following points:

❑ Suppose that, as part of a strategy for sustainability, a government decided to impose a resource tax, or an energy tax, or similar measure that would have the effect of raising the cost of resources. The effect of this would be to reduce the demand for investment capital for applications that generated increased demand for natural capital. However, it would increase the demand for investment capital for applications that would increase the efficiency with which natural capital could be converted into artificial capital, or that would otherwise decrease the demand for natural capital. The net effect on interest rates would depend on the relative proportions of these two types of investment in the economy at the time. If, for example, more money was invested in the first type of application, then, in the long-term, this money would be re-invested in the second type of application. However, in the short-term, the likely effect would be a net decrease in the demand for investment capital until a wider range of resource-saving investment options was developed. Such a net reduction in demand for investment capital would lead directly to lower interest rates.

❑ If technological progress made a more valuable use of a resource possible, the real value of that resource would increase. This would lead to increased returns on capital invested in the relevant sectors and in systems for the management of that resource.

❑ In some cases, and in some ways, artificial capital can be an effective substitute for natural capital. In these instances, investment in that form of artificial capital will tend to preserve the associated form of natural capital. If the effective value of that natural capital was then raised, either by technological developments or by some policy measure (such as conservation subsidies), this would tend to increase the value of the artificial capital investment. This would also lead to increased returns on capital invested in the relevant sectors. Such developments would tend to increase the prevailing interest rate.

❑ There would clearly be significant effects on investment patterns if discount rates were lowered or abolished. Any lowering of discount rates would have different effects on different groups of investors. Those investors for whom the investment per se is the priority and who have to justify investment against threshold discount rates would be able to increase investments. It is likely that this would enable, for example, additional positive 'ethical' investment which could have relatively low or long-term returns. However, current indications are that these investors are still a minority. 'Ethical' funds probably total less than 10 per cent of investment funds, and the majority of these funds are invested negatively (where certain opportunities are avoided on ethical grounds) rather than positively (where certain opportunities are actively favoured). The greater part of investment funds are currently managed with relatively few extraneous criteria. The effect on these investors, for whom the priority is to maximise the rate of return and minimise opportunity costs, would be more complex. For this group, any lowering of discount rates could have the effect of either increasing or decreasing investment, depending on the opportunity cost of the capital. If all investment became financially unattractive the relative value of present consumption would increase, so investments would decrease. However, if natural capital values, for example, were to rise in real terms, and were anticipated to continue rising at similar or higher rates, then investment in technologies that improved the efficiency of resource conversion and use would still demonstrate real rates of return.

Appropriate uses of discounting

While there is some agreement that there are a number of problems and anomalies associated with discounting, there is no consensus yet as to whether the principle should be reformed or abolished.[72] Some of the positions are as follows:

❏ The use of discounting is justifiable, provided that costs and benefits, and possibly 'sustainability constraints' are included. Of course, there are some practical problems in defining sustainability constraints. It would inhibit most forms of development if they were to be defined in terms of maintaining a constant stock of natural capital, which suggests that some form of substitution of artificial capital for natural capital should be allowed. However, it is likely that not all forms and functions of natural capital can or would be adequately substituted (see p 94). This would leave a need for compensating projects, but then the cost of these would have to be brought in to the same cost-benefit analysis.

❏ The use of discounting is justifiable for certain classes of project appraisal, but not necessarily for all. For example, public interest or other desirable long-term investment should not have to compete under the same discount regime. Of course, this approach could only function with clear guidelines as to which investments justified a lower discount rate.

❏ The use of discounting is not compatible with sustainability. Of course, sustainability would not axiomatically follow from the abandoning of discounting. There is no certainty that current or past rates of resource depletion, for example, are or were sustainable, irrespective of future rates. There are a number of models that are not based on the use of discounting (such as forest rent models), or are based on modified discount theory (where generational boundaries, for example, are used as constraints), or on environmental assessments (where environmental limits are used as constraints), or on some combination of environmental assessments and valuation techniques (such as cost-effectiveness analysis and various subjective scoring methods). All of these have certain inherent problems. A non-monetary approach does avoid the relatively extreme loss of data associated with a single numeraire and the necessity to collapse the various diverse reasons for discounting into a single discount rate. However, most models of this type, such as environmental impact assessments (EIAs), do not adequately distinguish between the different grades of significance of the different dimensions, nor do they make possible trade-offs explicit.

This indicates that both monetary and non-monetary models may be of use in an integrated approach. Monetary models will continue to be appropriate at the project level, but non-monetary models would be appropriate at system or policy level, which is where aggregate environmental impacts of various projects will become apparent.

Positional analysis, which is reviewed in detail in Chapter 13, is an approach that can incorporate cash revenues (discounted or not) and changes in non-monetary values. The advantage of positional analysis over standard EIAs is that grades of significance can be explicitly assigned and trade-offs clearly identified in the preparation of ideal profiles, against which all actual option profiles can be differentiated.

9

The Levels of Sustainability

There are important questions as to whether it is necessary or desirable to think that every country or region should be strictly sustainable, or whether every country is potentially sustainable. The concept of the nation state has real and important social, political and economic consequences. It has, for example, very significant effects for the way in which resources are allocated and used. The biological and geological processes that shape the Earth and which result in the distribution of both renewable and non-renewable natural resources, however, do not observe political frontiers. Given the very varied size and location of countries, the uneven distribution of resources and population, the disparate stages of development of the various nations and the various historical events that have resulted in current borders, it is inherently unlikely that all countries could be equally sustainable.

Following from this, it becomes apparent that one has to define the level at which sustainability is to be achieved. At any one level of analysis, whether it be an individual, a firm, a region, a country, continent or economic bloc, some factors are exogenous. For example, imports and exports from the unit of analysis may represent transfers of sustainability, pollution and other environmental effects may flow both to and from the unit, resource prices may be different inside and outside the unit (and may therefore affect flows of these resources), and so on. Furthermore, many significant flows of trade are controlled from outwith any single unit of analysis. For example, nations such as Japan, the UK, or Switzerland, with international financial operations and extensive holdings in primary and secondary sector companies in foreign countries, effectively control flows of natural capital in those countries. Only at the global level are all such factors endogenous. Thus sustainability must ultimately be determined within a global system. Of course, this poses some practical problems. Given the current lack of an effective global response or the structures and mechanisms for implementing any significant change at that level, we are faced with the task of deciding how to go about developing towards sustainability within existing nations such as the UK.

Systems and projects

It is necessary to break this problem down into two stages, which can be thought of as the *system* level and the *project* level.[73] This is because the important factor for a system as a whole is the total scale and rate of resource depletion, which can only be measured at the system level. Selective and partial controls at lower levels could simply displace effects to other parts of the same system. If the US, for example, were to halve its consumption of oil, the global benefit would be largely lost if other countries took advantage of the fall in price that would accompany the reduction to increase their consumption commensurately.

In principle, this overall control could be achieved in a number of ways, such as the deployment of appropriate economic instruments which would have the effect of driving up the price of a resource, or direct regulation to control or prohibit the use of certain resources. Either of these approaches would induce conservation and efficiency measures in every related decision made in every project contained within that system. This control could be flexible and sophisticated if necessary. If ecological thresholds were known to be a key factor then prices could be set to increase exponentially and to rise towards infinity as this threshold was approached clearly this approach depends on considerable confidence in knowledge of system thresholds, and it would be prudent to allow for suitable margins of error and contingencies. Where thresholds are not known, as is often the case with ecological limits, they would have to be estimated, which could be a problematic and contentious process, but which need not prevent the careful and considered application of this strategy.

An example of a system-level sustainability policy application might be a carbon tax or an energy tax. Such a tax would have the effect of automatically weighting every decision taken within the system towards greater economies and efficiencies in the use of energy.

The real problem arises when individual projects must be assessed in terms of their contribution towards overall sustainability. Any one project will only address a small part of the system that is to be sustained, making accurate and appropriate assessment difficult.

It would be possible, for example, for a given project to result in decreased welfare at a local or immediate level, yet to contribute towards increased welfare at a wider level or at some future time. The development of a wind farm in a scenic area might result in an immediate loss of local amenity. However, if the construction of the wind farm actually permitted the closure of a fossil fuel-burning power station, and thereby displaced carbon emissions, there might be a gain in national or long-term amenity. If the wind farm, however, contributed only to an existing generating overcapacity and thereby weakened the case for energy conservation, did not lead to the closure of a fossil fuel burning power station, and did not therefore displace carbon emissions, there might be little compensating gain for the loss of amenity.

As another example, imagine that society adopted a system-level constraint to the effect that natural capital depletion, or other environmental damage, should be zero with non-renewable capital and zero or negative with renewable capital. Of course, with non-renewable capital, this would simply mean an absolute restriction on use. In the case of renewable capital, however, it would be very restrictive to apply this constraint to each individual project. It would also be unnecessary, given that the critical factor is aggregate positive or negative changes in the natural capital balance. This can only be seen at a strategic level at which it is possible to monitor and calculate the net direction of such change.

The conclusion is that all projects must be assessed in the context of a strategic system-level programme, within which costs and benefits can actually be identified, smoothed out over time, and integrated.

The role of determining system-level programmes, controlling macroeconomic parameters, and calculating changes in the net natural capital position is one that, given current modes of human social organisation, could only be carried out by governments and their agencies or by intergovernmental institutions. There are a number of options for integrating market operations into this process. For example, a central government agency could confine its role to market correction and regulation. Alternatively, the market could be permitted less constraint if the central agency itself provided an environmentally-enhancing (but probably 'uneconomic') project that compensated for environmentally damaging market operations. This would not be a simple solution, as it would then become necessary to determine what would constitute adequate compensation and to decide on the optimal deployment of the resources available for compensatory projects.

A relatively simple example of a system-level market correction would be the application of a resource-use tax, or an energy tax, or conservation incentives. Almost any desired resource depletion rate could be achieved with the application of the appropriate level of revenue-positive, revenue-neutral, or revenue-negative fiscal measures, although it is probable that there would be some sectors which would demonstrate greater inelasticity than others.

It is significant that a number of governments apply measures that are effectively resource depletion incentives, such as tax concessions for mineral extraction operations, underpriced logging permits, and so on. These are instances of system-level market controls which will tend to operate against sustainability. In these cases, governments may be operating with a higher discount rate than the private sector, perhaps influenced by short-term political factors. It would be relatively simple in principle to replace these with measures that had the reverse effect, though in practice it would be likely to be very politically difficult. Such changes tend to be resisted by those who benefit from the status quo.

The strength of the measures that may be required will depend on the current position and policies of a society. Given that some societies are likely to be further away from sustainability than others (see Chapter 4 and p 90), stronger measures or a longer period of adjustment may be needed in some instances.

Capital flow

The issues reviewed in Chapter 7 have a number of implications. One of the practical problems that arises is how to establish a 'sustainable project' in the context of the current economic order.

In the experimental or developmental phase of a project a standard pattern of investment results in a net negative (inward) flow of capital. This is not indefinitely economically sustainable, as a sustainable project cannot depend forever on resources drawn from elsewhere. Thus a mature sustainable project is likely to have a neutral (the self-reliance scenario) or modestly positive (outward) net capital flow, averaged over an appropriate period of time. While it is clear that a development option is not indefinitely sustainable if it depends on continuing flows of capital (which could be in the form of funds, personnel, or other forms of support) from outside, it is also possible that an excessive flow of capital outward, in the taking of profits, for example, could engender an inability to invest adequately in future development, or be symptomatic of a failure to allow adequately for depreciation or a failure to internalise the 'true' costs of the operation.

More fundamentally, if the prevailing economic order is founded on unsustainable practices, then it may be impossible for sustainable projects to be economically viable when competing with projects that are allowed large and unacknowledged subsidies. For example, a competing operation might have poor emission standards and thereby externalise some of the costs of production, or in other ways consume the 'free gifts' of nature without allowing for the cost of replacing these and repairing the damage done. This represents unfair and unequal competition for a project that internalises these costs, as profits are maximised, by definition, by the externalisation of costs. Anomalies of this nature can be seen in a number of areas today. For example, rail-based transport systems are generally environmentally preferable to road-based systems. This is because they are more energy efficient (due to larger passenger and freight capacity and lower friction resistance), incur lower opportunity costs in terms of the land and space required (a rail system requires a far narrower corridor of land than a road system to move the same amount of people and freight), can utilise a greater variety of primary fuel sources (electric trains can be powered by nuclear, fossil fuel or wind-turbine generated electricity), and have longer unit lives (trains are built to last longer than cars). However, between 1952 and 1988 UK rail passenger kilometres remained relatively constant, at about 100 billion passenger kilometres per year (pk/y), while car passenger kilometres increased from 180 to 600 billion pk/y. These discrepancies have arisen for a number of reasons. One factor is the relatively heavy indirect subsidies paid to car users. Road users do not pay the full cost of constructing and maintaining a road network, for example, especially in rural areas. Low fuel costs and poor environmental standards represent especially large subsidies, where resources are effectively transferred from the common environment, the wider community, and future generations. True rail/road system comparisons cannot be made until such subsidies are identified and stripped out.

Similarly, it might be difficult for any country to maintain high environmental standards for forestry, for example, if this pushed the price of that country's timber significantly above the world market price. However, the world market price for timber may be being held at an artificially low level due to the willingness of some less enlightened, poorer, or more corrupt nation to log its own forests to destruction.

Anomalies of this kind would only be resolved if the entire global economy was reformed on the principles of sustainability, so that all economic operations were obliged to absorb energy, resource, and emission costs that are not currently borne. In this situation all projects would compete economically on an equal basis. Such a global macroeconomic set of rules does not currently exist.

There are, however, several developments that might eventually allow more sustainable projects to compete on a more equal basis. These include the following:

❑ The growing consumer preferences for products and services produced and delivered to higher environmental and ethical standards.
❑ The possible eventual introduction of market corrections, such as carbon taxes, tradable pollution permits or other measures towards a more direct attribution of environmental costs.
❑ The introduction of legislation and international accords to prohibit certain particularly damaging products or practices.

Until such developments occur, it may be necessary to subsidise those projects that absorb their energy, resource, and pollution costs. Another option would be to develop projects that were environmentally enhancing, as compensation for other projects that were environmentally damaging.

10

Financing the Transition to Sustainability

Financing reconstruction

It is a reasonable assumption that a society that prioritises the production of essential goods, produces these with ever-cleaner technology, and operates to increasingly high standards of energy and resource-use efficiency will make a greater contribution to global sustainability than one that fails on any or all of these grounds. A transition to the foregoing from the current UK position is likely to require some degree of reconstruction of many sectors of economic activity, the industrial base, transport infrastructure and so on.

This is a key problem in devising strategies for a transition to sustainability. The process of transition itself has obvious financial implications. Significant amounts of capital would be required. A transition to renewable energy sources, for example, would require an expensive restructuring of the generating and supply industries, because the renewable energy sources have different characteristics, such as relatively low load factors. Thus the transition would be expensive, although the more efficient use of energy flows and resources might well reduce costs in the long run.

One option is that governments could increase taxes and invest the revenue in relevant infrastructure projects. In a market or mixed economy, however, large parts of the productive sectors are in private hands. These sectors must also be able to raise the necessary capital. This could be delivered in the form of government grants, but this would be an additional burden on the taxpayer. It might also look like a subsidy to precisely those firms that had failed to invest in environmentally-clean technology. The additional cost of developing and implementing cleaner technology, for example, would be high for those sectors and firms that had recently invested in manufacturing plant built to operate to inadequate standards, or who had paid out their profits as dividends rather than re-investing them, but low for those socially and environmentally-aware operators that had already moved in this direction, or who were just fortunate with the timing of their investment decision. This might not be generally seen as fair or just.

Private firms are, of course, able to raise capital in a number of ways, some of them extremely ingenious. There are many sophisticated financial instruments available – bank loans, bonds, debentures and different kinds of debt, ordinary shares, preference shares and other different kinds of equity and so on. In broad terms, however, they can either reinvest their own profits, borrow the money, or sell shares to investors.

There are a number of features of the UK markets in money and shares that could make it difficult for firms to raise the capital required to reconstruct their operations and to put them on a more sustainable basis. This section therefore reviews the operations of the financial markets, with reference to the functional and structural aspects that could form a significant impediment to a transition to a more sustainable way of life.

Financing investment from profits

It may be possible in some instances to finance the necessary investment from profits. This could be a financially attractive solution provided that the opportunity cost of the capital was not too great. If the penalties levied for environmental breaches, for example, were light, then the cost of a company's failure to meet environmental standards would be low. It might be more financially advantageous to pay the fines while leaving profits accruing interest in the bank or otherwise invested. If environmental standards and associated sanctions were high, the cost of non-compliance would be high. This would have the effect of reducing any opportunity cost disparity, so that there would be a greater incentive to invest in actual improvements.

There are, however, two factors that make it unlikely that many firms could finance the investment entirely from profits. First, many companies are under pressure to maintain high rates of dividends to shareholders (see p 158). This reduces the funds that would otherwise be available for reinvestment. This problem could be partly overcome by the introduction of appropriate regulatory measures. Raising environmental standards, for example, would oblige some companies to reinvest in new technologies to the point where the company would have to reduce its dividend, which would make those companies that already have relatively high environmental standards look comparatively attractive to investors.

Secondly, there is the level of expenditure that would be required, in some cases, to make a transition to cleaner technologies. For some firms, the expenditure required would be such that they could not be financed from profits. In some instances, this will be due to fluctuations in the fortunes of the firm or sector, so that a firm or sector in long-term decline, or that has just experienced several years of poor profit levels, would not be in a position to make the investment required. However, it is also more likely to be true of those companies that generally have low rates of reinvestment, as these companies are likely to have a greater 'investment deficit'.

There is a possible exception to this, which is that in sectors that are developing particularly rapidly, or where significant change in the legislative environment is imminent, delayed investment might allow a company to avoid investment mistakes. It could also allow a company to skip a generation in terms of the technology required. This approach may make sense in a strategic context, but should not be confused with simple failure to invest per se.

Financing investment from equities or borrowing

Companies that cannot finance the necessary investment out of current revenues will have to raise the finance elsewhere. There are two main sources of finance for any company proposing a major reinvestment. First, a company may issue equities. Second, a company may borrow funds, either from banks or by issuing bonds.

However, the current modes of operation of the UK money markets, changing attitudes to risk, and the current ratio between equities and borrowing may represent problems for any proposed financing of a transition to sustainability.

Equities

One of the main advantages of raising finance through issuing shares, as opposed to borrowing from banks or issuing bonds, is that dividends can be varied. When in recession, or at the low point of an economic cycle, dividends can be reduced. Interest rates levied by banks, however, remain the same. If the recession is particularly deep or prolonged, and if a substantial fraction of the bank's customer base fails or otherwise defaults, interest rates may even rise. Similarly, bonds must be redeemed by due date, irrespective of the changing circumstances of the company. Thus issuing shares can be a relatively attractive way to raise funds.

However, raising money in this way has a number of other important implications for the company. One very important factor is the effect of the secondary and tertiary markets. When a company launches a share issue, it does so into the primary market. This share launch into the primary market raises cash for the company. These shares are then further traded between institutional, corporate, and individual shareholders in the secondary market, which is much larger in terms of the number and nominal value of the transactions than the primary market, while options to buy shares are then traded in the tertiary market, which in many sectors is now larger than the secondary market. This additional trading in the secondary and tertiary markets does not raise cash for the company. It does, however, affect the market value of the company. This in turn determines the 'gearing ratio' of the company's debt to its net worth, which affects the credit rating of the company and hence its ability to borrow further money or attract further investment. These factors affect the perceived strength of a company, and the extent to which a company then becomes vulnerable to takeover.

However, the use of equities to raise funds and the role of the secondary market per se are not necessarily problematic. They only become problematic in the context of the current UK market culture. Any market, being a social phenomenon, is underpinned by cultural and psychological factors such as perceptions, beliefs and expectations, which are reified in the form of institutions and legal frameworks. In the current UK context, although the shareholders of a company are its owners, many shareholders do not in practice behave as if that status conferred the obligations associated with various other forms of ownership. Many do not perceive themselves as holding any responsibilities to the company, for example, and behave as if they were a company's creditors, or rentiers. Thus shareholding is widely seen as a unilateral obligation assumed by the company, in exchange for cash, to pay to whoever currently holds the shares such dividends as are due.

Should a company be made responsible for cleaning up not just current but also its historical pollution, for example, the first response of many shareholders would be to sell their shares in anticipation of reduced dividends. This would not be a helpful response from the point of view of the company, society or the environment, but would be rational behaviour by current market cultural norms. More fundamentally, any company which truly internalised its environmental costs and thus behaved in a properly responsible manner could not compete, at least on price terms, with an unscrupulous rival that externalised as many of its costs as possible. Thus the investment funds would flow into the unscrupulous company rather than the responsible one.

Similarly, should a company decide to anticipate higher environmental standards, and invest profits in new technologies and restructuring programmes instead of paying them out as dividends, the response of many shareholders would be to sell their shares. The investment strategy might work to the company's advantage in the long term, but many shareholders would take the years of higher dividends elsewhere, and consider moving back when the investments started to show returns.

In general, should a company not maintain a high level of dividends, many shareholders will normally sell those shares and move funds into shares that offer better rates of return. This might not matter too much if shareholding were widespread and dilute, as people would not necessarily have the same perspective of a company's position and actions. However, more than half of the equity in all UK companies is owned by some 60 financial institutions such as pension funds, so that the decisions taken by a small group of investment analysts tend to have a great deal of weight. A decision by one or two major market operators can, therefore, depress a company's share price. A sharp fall in share value could mean a less favourable gearing ratio of debt to net worth, and to the loss of a company's good credit rating.

This can be exacerbated by market volatility. Heavy selling of shares is often self-perpetuating for a certain period, as relatively slight movements can be amplified by rumour and panic. Market traders do not wish to get left with devalued stock and heavy losses, so may try and sell early if they anticipate further falls in share prices, thereby strengthening the trend. In this way, a company can suffer a greater reduction in share price than is strictly justifiable in terms of reduced share yield. Many traders now use computers to monitor share prices, with automated buying and selling being triggered at set points, which may reinforce this particular market phenomenon.

Thus the operations of the secondary markets, in the context of the current UK shareholder culture, place companies under pressure to maintain dividends. For example, when company profits rose as a proportion of GDP during the early 1980s dividends rose too. However, when company profits fell in the early 1990s, dividends remained high. In 1991 UK companies only reduced their dividends on aggregate by 1.2 per cent, in spite of the prolonged downturn in the economy. In 1992 dividends stood at 4.3 per cent of GDP, compared with 2.0 per cent in 1979.

This situation has several consequences. First, it tends to make management more risk-averse, and more willing to sacrifice long-term investment in favour of maintaining a high level of dividends. As the remuneration of management is often linked directly to share prices, directors often prioritise actions that will affect share prices to the detriment of other legitimate corporate goals.

Secondly, it places a premium in the secondary markets on those companies whose dividends are safe and likely to continue rising. This is likely to include companies that are sheltered from international competition, that have secure markets, and that take few risks with new technology. This is less likely to include companies that are exposed to greater international competition, that have fragmented, or rapidly evolving, or otherwise insecure markets, and that are taking greater risks with innovation.

In general this will tend to favour, for example, food retailers over engineering firms. This is the kind of factor that could make it difficult to raise the funds for the significant investment in engineering that would be necessary to reconstruct the industrial base.[74]

Borrowing and credit

Companies can also borrow money in various ways, often from banks or investment funds. Although the terms and conditions may be more complex than for a personal mortgage, the general principles are essentially the same. A sum is borrowed, and repaid within a certain time with a certain rate of interest.

A banking system has several important advantages, in a wider context, over the sale of equities as the main source of funds for long-term investment. One important advantage is that banks can create credit. Essentially, this works as follows. Banks lend against savings. Much of the money that they lend remains on at least temporary deposit with the bank, or is spent and then redeposited with the bank. This means that further loans can be issued against these new deposits. In countries with a large number of small banks (such as the US), as opposed to a small number of large banks (such as the UK), this initial redepositing is less likely to be with the same bank, but this does not matter in this instance as a banking system serves the same function in this regard as a monopoly bank. This process can cycle a number of times, which enables banks to create credit which may be many times the volume of the original saving. After various historical financial disasters, the amount of credit that a bank is allowed to create in this way is now controlled by the reserve requirement imposed by central banks.[75] For example, a reserve requirement of 25 per cent allows a bank (or bank system) to create £4 credit for every £1 deposited, while a reserve requirement of 10 per cent allows a bank to create £10 credit for every £1 deposited.

This process allows a banking system to access a greater percentage of the savings and circulating credit in an economy than a stock market. The ultimate source of this circulating credit is the central bank, which determines the reserve requirements. Of course, this circulating credit is withdrawn, in exactly the same ratio, when central banks sell (for example) government bonds. The cash withdrawn from the system removes the same multiple of credit as was created.[66]

This ability of banks to create credit has several different implications for the kind of investment strategy that would be required to finance a reconstruction of industry. It depends on whether the credits would be used in practice primarily for investment or for consumption.

If the credits created were used to finance investment, this would encourage innovation and development. An appropriate regulatory or planning framework could then be used to direct such investment into improving, for example, energy and resource-use efficiency.

Investment credits also tend to fuel long-term growth in the economy, and particularly rapid growth occurs in those economies, such as the Japanese economy, where investment credits exceed the total of national savings. This means that investment is being funded partly out of savings and partly out of credit-created debt. Further growth is then generated by inflows of foreign investment capital, which is attracted by the internally-generated growth. Thus an investment-led strategy, with funds channelled into the next generation of energy-efficient, resource-efficient and cleaner technologies, would have certain additional advantages.

If, however, the credits were used to finance consumption, this would tend to fuel inflation, as credit for consumption generally increases demand. However, this demand would not necessarily mean propor-tionate increases in domestic supply. If this situation were to occur in an economy which for some reason, such as long-term failure to reinvest, does not have sufficient manufacturing capacity to meet this increased demand, the consequence would be an increase in imports and a deterioration of the balance of payments position.[76]

Unfortunately, the latter appears to be a closer description of the current UK economic situation. This has arisen for various reasons, some political and some related to the nature of the market. Part of the problem is that the kind of short-term bank lending used to finance consumption has been more profitable than the kind of long-term lending needed to finance investment (see p 161). This ratio would have to be reduced or reversed if significant amounts of capital are going to be released for investment. It is difficult to see how this could be achieved without some government intervention in the market, and possibly the introduction of additional forms of financial regulation. Given the complex, fluid and increasingly international nature of the money markets, this would probably require the coordinated use of a number of policy instruments both nationally and internationally.

There would be one obvious temporary problem with the use of banks to finance a move to cleaner technology. Banks usually require security against a loan. Traditionally, this is land or capital equipment. In some cases, however, land left in a contaminated state after industrial operations may prove to be a negative asset, as the cost of cleaning the land may be greater than the value of the land. In any reconstruction of the more polluting sectors of UK industry, it might therefore be necessary to find some new form of security.

Conclusion

In conclusion, a transition to a more sustainable way of life is likely to require significant investment, and this capital will have to be raised. It will be necessary to find finance for inherently long-term investments, such as the reconstruction of polluting industries, the development of energy-efficient transport systems, the cleaning-up of historical pollution and so on. If these funds are to come, at least in part, from the market, then the priorities of the current UK market must be changed so that long-term investments are not penalised.

This suggests that there would have to be a number of changes in the function and structure of the market itself. These might include the use of financial incentives and penalties to encourage long-term investment and reduce the short-termism of the market, moves to link the remuneration of managers to factors other than share price alone, a redefinition of the legal responsibilities of directors to encompass wider concerns than the maximisation of profits, and measures to reduce the vulnerability of companies to hostile takeover when share prices are temporarily depressed by major programmes of investment or environmentally responsible behaviour such as clean-up operations.

Another useful change might be to de-emphasise the role of equities in raising capital, and to augment the role of the banking system, in conjunction with the introduction of measures to ensure that the additional credit that would be created in this way was used for investment rather than consumption, and that the investment was towards greater energy- and resource-use efficiency, cleaner technology and so on.

There are a number of developments, however, such as the growth of the ethical investment funds, employee share-ownership schemes, share-based cooperatives and organisations like Traidcraft, which have successful issues of shares that bear no dividends but which are bought by people who support the aims of the organisation, which support the idea that the use of equity issues to raise funds need not necessarily be a problem in itself. Changes in the current UK market culture would be helpful in any case, and might be sufficient to remove some of the problems detailed earlier in this section.

Another important change, therefore, would be to reform the legal rights and obligations of shareholders. The Trustees Investment Act, for example, places trustees under a legal obligation to maximise the returns to their investors. This would have to be amended if more long-term, socially and environmentally responsible investment was to be encouraged, given that any environmental cost successfully externalised adds directly to a company's profit margin.

More generally, programmes to raise the awareness amongst shareholders of the wider social and environmental implications of corporate operations could be used to encourage a greater sense of responsibility to the company and for its operations. Long-term shareholding could be encouraged with incentives, which need not invariably be financial.

In systems terms, the need is to construct the missing information feedback loops between corporate operations and environmental consequences, and between companies and the shareholder-owners of companies, so that those responsible for particular actions become aware of and directly affected by the consequences of those actions. Any move towards greater transparency and accountability in the markets for either money or shares would be helpful in this regard.

11

Economic Development and the Environment

This chapter considers economies as complex adaptive systems, reviews some of the historical events that helped to shape contemporary economic theories and structures, then considers which economic model offers the best prospect of being able to generate and maintain the kind of flexibility referred to in the discussion of a systems model of sustainability on p 38. Our preferred model is an appropriately regulated market system, with a role for the state in correcting market failures and in strategic planning and investment. There are a number of different market systems in the world today, some of which are currently closer in certain key respects to our preferred model, and which we believe might therefore be better placed to make a transition to more sustainable modes of operation. Should environmental constraints become more pressing, we would anticipate that the countries that are better placed in this regard should be able to develop or increase a relative economic advantage in a transition to more sustainable practices.

Economies considered as complex adaptive systems

One of the most important challenges to any economic theory is to explain the process of economic change and growth. The historical record of economic change and growth since the start of the industrial revolution is rich and complex, and allows for more than one possible explanation.

The industrial revolution – the start of the general mechanisation, industrialisation and the increased capital intensity of production, and the associated step-change increase in the rate of generation and development of new technologies – led to a period of sustained growth in productivity and per capita income. Economic growth rates started rising in the last decades of the eighteenth century, and increased throughout the nineteenth century into the early decades of the twentieth century. The period from 1945 to 1970 saw even more rapid growth rates in many countries, followed by a significant reduction in growth rates from 1970 onwards in many of the advanced industrial nations. There have been a number of distinct eras, however, in this broad sweep of change and development since the late eighteenth century, each characterised by different leading technologies with their associated institutional and financial arrangements. The rates have also varied between nations and regions of the globe, with the UK showing the highest growth rate until the late nineteenth century, then being overtaken by the US and – in some fields – Germany. Japan developed very strong growth rates after the 1950s, which slowed recently, while the Pacific Rim nations have now taken the lead in many sectors. The main engines of growth have been progressively superseded and replaced, therefore, but the process of growth has, in general, continued throughout.

There are a number of theories of economic change and growth, which fall broadly into three groups. One group, which is rooted in a neoclassical perspective, consists of those theories that are principally concerned with the relationship between growth in inputs (such as the growth in capital per worker) and growth in outputs (such as the growth in output per worker). This school employs a relatively mechanistic model of change, tends to focus on the forces that move the economy towards equilibrium, and is not, in general, concerned with the details of the process through which technological innovation occurs.

A second group, which is informed by an institutional perspective, is more concerned with the role played by institutions and structures in the national economy in shaping economic expectations and behaviour.

The third group consists of the evolutionary theories of economic growth. This group is concerned with the details of the process through which technological innovation occurs, and with the forces that upset equilibrium as much as with the forces that move the economy towards equilibrium. The focus is usually on the individual firm, with the growth of output and capital at the macroeconomic level seen as the consequences of the microeconomic generation and dissemination of technologies. The second and third groups overlap in a common belief as to the importance of institutional structures.

There are a number of reasons why the authors believe that the evolutionary theories of economic change and growth will, ultimately, offer a more satisfactory explanation than the mechanistic model. It is important to emphasise that the evolutionary approach must offer more than just a sophisticated biological metaphor if it is to fulfill this promise, especially given that a number of economists (including a number of neoclassical authorities) have already drawn on biological metaphors in their expositions.

The recent development of the school of evolutionary economics, with roots in the school of institutional economics, is the result of a more formal attempt to see whether the principles of evolution can be applied more generally to explain the processes of change in systems other than biological systems. Evolutionary economics, therefore, invokes general principles concerning systems to make inferences about interactions in the domain of economics. From this evolutionary perspective, large-scale features of economies emerge from the detailed interactions of individuals, firms and sectors of industry. The main focus of this school to date has largely been on such macroeconomic questions as the differences that characterise national economies, the rate and route of diffusion of technologies and the nature of economic change, and such microeconomic questions as the detailed dynamics of the processes through which firms and other economic operators arise, grow and compete.

The emerging literature suggests that there are a number of ways in which biological systems appear to have certain formal analogues in economic systems, and economic systems manifest certain abstract structural similarities to biological systems.[77] In evolutionary economics, the firm, rather than the organism, is the effective unit, but the problems of survival and growth (or reproduction) bear certain fundamental similarities to those in biological systems.

Organisational knowledge – the shared knowledge, learning and culture of an organisation – plays a role similar to that of biological homeostasis in maintaining stable patterns of production and behaviour while individual human members of the organisation come and go. Such knowledge largely defines an organisation's opportunities and constraints. The knowledge base, in economic evolution, is the analogue of genetic information in biological evolution. It generates structural and behavioural stability for the organisation, while the 'selection pressures' of the competitive marketplace eliminate unsuccessful strategies.

Adaptive change in the knowledge base is therefore the key to the transformation of these opportunities and constraints, and hence to the chances of survival and growth. Some change is relatively easy, however, while other structural features of the organisation – such as established routines or decision-making procedures – tend to be much harder to change.

There are similar opportunities and constraints in biological evolution. During the evolution of the mammals from the therapsids (primitive mammal-like reptiles that lived some 200 million years ago), for example, a bone that forms part of the jaw in reptiles was transformed into a component of the mammalian ear – the first mammals had an intermediate structure with two jaw hinges, the old reptilian hinge and the new mammalian hinge operating side-by-side. This appears to have been a relatively straightforward transformation, as each step probably offered a consistent advantage in the form of improved hearing and stronger jaws. In addition, the reptilian version of the bones already played some part in hearing, and the bones did not have far to move in order to become part of an ear.[78] These factors created a viable transformation pathway. More awkward changes, possibly incurring some disadvantages as well as advantages, can and do occur in nature, but some potential changes are clearly more likely – in a given context – than others.

The process of technological development and transfer exhibits dynamics similar to those in biological systems, as the flow of new technologies from development into production and use appears to be shaped and constrained by the prior structure of the organisation. A given technological paradigm, once established, then yields a basic configuration of artifacts. It imposes an overall dominant design. Almost all cars, for example, share the solutions to certain technical problems, such as the process by which energy in the form of heat is converted into mechanical movement. The competition amongst car manufacturers depends, in part, on their success in implementing these solutions.

This too has a biological counterpart. The process of evolution arrives at biological paradigms, which then dominate and shape further development. It has recently emerged that the basic problem of coding and controlling body forms, for example, was first 'solved' with a set of genes called the *Hox* genes. It now appears that the body plans of all animals are coded and controlled via variants on the descendants of the same set of *Hox* genes, rather than by using new and different sets of genes for each species. Particular body plans arise, compete, are more or less successful, and are perpetuated or eliminated as a result.[78]

Evolutionary economic models draw directly from models of biological evolution. They assume, in general, that firms that develop more efficient production processes and more profitable products will grow faster than their competitors, but that the competitors may then seek to innovate or imitate in order to improve their efficiency, that inefficient firms fail and die, and that a flow of new firms into the sector will change the environment by introducing more variation in the range of competing technologies and behaviours. Such models also assume that the development of a firm is constrained by existing characteristics (their organisational structure and knowledge), which limit the kinds of outcomes that can occur, but that all change is accompanied by some degree of chance or random variation. Such variation can generate new and unforeseen possibilities, some of which will be neutral in terms of economic advantage, at least in the current circumstances, some of which will be damaging, and some of which will open up new opportunities.

Evolutionary economic theory, unlike neoclassical theory, uses the same models of human motivation and behaviour as the rest of the social sciences. Thus evolutionary economic theory does not assume, as neo-classical economic theory does, that what firms do is optimal. It assumes that firms will, in general, try to survive and make profits, but that their decisions are not necessarily rational, value-free, or made on the basis of perfect information. It assumes that decisions are, in practice, influenced by the existing organisational knowledge and culture. It also assumes that firms will adapt as their competitive environment changes, as their internal turnover of staff generates new behavioural rules, and with the development of new technologies, which will all affect the range of their operational constraints and opportunities.

One of the general features of most conventional (non-evolutionary) theories of economic change and growth is that their predictions tend to be reliable only during periods of relatively smooth and continuing growth. Sharp changes in the pace and pattern of economic growth, caused, for example, by the successful development of some new technology, appear anomalous. One of the important advantages offered by evolutionary economic theory is that it readily lends itself to nonlinear dynamic modelling, in which such occurrences are to be expected.

Most conventional economic models suppose that economies can arrive at various points of equilibrium, at which the forces that determine an outcome will be in balance. The matching of supply and demand, the law of diminishing returns, and movement on the production–possibility frontier, for example, are thought to behave in this way. Much conventional debate has centred on the question of whether economies function most efficiently when left to find these points of balance, and thereby generate the most wealth, or whether governments should take corrective or counter–cyclical measures to try to achieve different outcomes.

In reality, however, economies do not appear to exist in static equilibrium. They change both quantitatively and qualitatively; they fluctuate, expand, and collapse. This kind of behaviour is more readily understood within evolutionary economic theory, or from a general systems perspective, although it is currently impossible to be certain that economies (being human systems, and so subject to self-fulfilling predictions) are even as predictable or regular as chaotic physical systems.[79]

Ormerod, for example, has presented a compelling argument that the classic equilibrium model of an economy does not describe reality, that a given economy will have multiple equilibrium points, which appear to function in this regard as system attractors, and that particular changes in market conditions can dislodge the economy from an oscillation around one equilibrium point and move it into the orbit of another.[16]

This is the kind of behaviour that it is hard to explain within the neoclassical economic paradigm, but which would, broadly, be expected of a complex dynamic system. This work by Ormerod, amongst others, suggests a new agenda for macroeconomic research, and that the recent insights developed in complex adaptive system research could usefully be extended to economic analysis.

For example, many different kinds of systems have thresholds – points at which the behaviour of the system changes qualitatively. A particular event in such a system may seem small in isolation, but may precipitate an apparently disproportionate effect if the system as a whole is poised near a threshold of transition into chaos or into a new balance. Thus one last straw can break the camel's back, or the fall of one pebble can start an avalanche, provided that the system is near a threshold. This suggests various possibilities for further research. It would be possible to examine, for example, whether some relatively minor economic event – perhaps a pessimistic forecast, or a rise in the price of a key commodity – might, in some circumstances, be sufficient to trigger a chain of events that leads to a loss of market confidence and consequent economic contraction.

Many different kinds of complex systems are also sensitive to the incremental results of the interactions of the elements of the system, and demonstrate positive and negative feedback loops. There is general agreement that economies demonstrate at least some of these system features. Negative feedback loops that serve to maintain balances between supply and demand, for example, are fairly well understood, and can be accommodated within the neoclassical economic paradigm. More controversially, however, there also appear to be positive feedback loops, that can generate such phenomena as increasing returns, and which can be less well accommodated within the neoclassical paradigm.[80]

Imagine a situation, for example, where there are two competing technologies, and where the customers do not have perfect information as to which technology they should adopt. There will be some penalty involved in making the wrong choice and adopting the unsuccessful technology. At the start, then, some customers will probably adopt one, some the other. Evolutionary economic theory suggests that many will use an informal probabilistic set of rules in making their choice. One rule, for example, might be to follow the majority by imitating friends, neighbours, or influential commentators in the media. Models of such cases indicate that the relative shares of the two competing technologies do arrive at a relatively stable final state, and that a relatively small advantage for one at the outset tends to translate into a dominant position at the end even if the unsuccessful alternative was an objectively better technology. Thus conflicts between competing engineering solutions, such as recording formats, for example, will usually end in one becoming dominant because even a slight advantage can translate into more machines and more material becoming available in that format, which will mutually reinforce and lead to further switching into that format. It is not hard to find instances of this effect, such as the conflict between the VHS and Betamax videotape formats, for example, which ended in the domination of a format which most believe to be technically inferior, or the emergence and domination of the QWERTY keyboard layout over the various rival solutions, even though the QWERTY layout was actually designed to be ergonomically poor and slow operators down (because the early keyboards were liable to jam when used at speed).[46]

Thus an imperceptible initial advantage can lead to market domination, in a way that the neoclassical theories of competition do not predict but which can be seen in a number of models of complex system behaviour. The existence of positive feedback loops means, in principle, that the events that give market advantage and eventual domination could be very slight and apparently arbitrary. The solution that emerges will not necessarily be the best even of those available at the time, which means that markets cannot always be expected to generate optimal outcomes in, for example, engineering terms. Furthermore, the existence of such apparently arbitrary effects means that the development pathway taken by an economy could become quite unpredictable, especially given the limits of current understanding of complex system dynamics.

It appears likely, in the light of current knowledge, that the accuracy of predictions of complex system behaviour will always fall off rapidly over time. While it may be possible to predict the weather, for example, for some years ahead in terms of broad patterns, it may only ever be possible to forecast in detailed and precise terms a few days ahead. This is because minute variations and inaccuracies at the outset can cause radically different outcomes. This is sometimes referred to as the butterfly effect, from the idea that the beating of a butterfly's wing in one hemisphere can change the weather in another.[81] Another item for further research, therefore, would be to examine whether this could be true of economic predictions too. At the same time, various observers have noted, for example, that predictions made more than a year in advance can be more accurate than short-term predictions, because certain variables are more stable over the long-term.[82]

If such further research suggests that economies are usefully viewed as complex adaptive systems, and if it then proves to be true that economies – as with other such complex systems – can only be modelled in broad terms, or for a relatively short time ahead, then the emphasis in government macroeconomic management would have to shift from the search for points of static equilibrium to dynamic system management, with a special emphasis on the management of information. The economy would have to be continuously monitored and evaluated, and this information fed back into a continuous decision-making process, so that the management policy could itself adapt and evolve. This would require a concurrent change in the political culture, from one in which changes in policy are seen as being inconsistent, and penalised, to one in which specific policy was allowed and expected to evolve within broad strategic and ethical parameters. As Galbraith[84] put it: 'official forecasting has an ineluctable tendency to error... the solution is not better forecasts but prompt and unapologetic accommodation to what exists, and prompt and unapologetic accommodation when that no longer exists.'

If, in the light of such further research, it also proves to be true that economies have multiple equilibrium points, the economic debate itself must move to a higher dimension. If there is no one 'natural' equilibrium point, and if it proves possible, in the light of such understanding, to move the economy to a different equilibrium point, then there would be a range of political and social decisions to be made as to which would be the most desirable equilibrium point. It might be possible, for example, for society to choose its preferred point of trade-off between levels of inflation and unemployment (if such a trade-off exists).

Furthermore, if economies do behave as complex adaptive systems, then it would follow that different economic systems could be expected to have different behavioural characteristics. Significant structural and functional differences, such as the extent to which an economy is regulated, the extent to which a government is prepared to take counter-cyclical measures, the institutional structures and policy frameworks present, and the socio-cultural understandings that underpin that economy, for example, would, in principle, impart different behavioural characteristics to an economic system. Economic systems with different internal structures would then be expected to respond in different ways to significant external perturbation or shocks. Ormerod, for example, has described the diverse ways in which different European economies with broadly similar rates of inflation responded to the shock of the 1973–74 oil price rises. The price rises led to a doubling of the rate of inflation in Italy and the UK, but a reduced rate of inflation in Germany. This Ormerod attributes to certain internal features of the German economy, in particular, the more general sense of consensus and mutual responsibility engendered by the German social market model.

Some of these points are considered further, in the context of managing a transition to sustainability, later in this chapter.

Industrialisation and associated change

The industrial and post-industrial era has done much to shape today's world. It has significantly altered the way of life of most of the peoples of the world, changed the nature of power, structured much of our way of life, and accelerated or caused some of the most pressing environmental problems.

It has also shaped most of our current economic theories. It is important to review briefly some of these wider implications of the process of industrialisation before going on to consider theories of the relationship between industrial and economic development and the environment.

Human activity has altered and in some instances damaged the environment for thousands of years. Environmental damage is not exclusively an industrial phenomenon, although industrialisation has accelerated a number of processes which cause environmental damage. The process of industrialisation does not occur in isolation, but both depends on and causes other social and economic transformations. This means that environmental costs associated with a particular pattern of development generally result from a combination of co-evolving factors, rather than the process of industrialisation per se. For example, the industrial revolution that started in Britain around 1750 was enabled partly by prior developments in agriculture and a transition to a market economy. These had the effect of breaking the links between labour and the land (which created the necessary workforce), and fostering the growth of the cities (which created the necessary markets).

Similarly, the depression of the 1870s and 1880s was the most important cause of the transition from earlier modes of political imperialism to economic imperialism, as predicted by Adam Smith. Non-industrial nations were developed as sources of raw material and cheap labour, usually via controlled operations that repatriated profits to the colonising industrial power, and as new markets for the products of industrial capitalism. This had the effect of destroying indigenous economic structures and replacing these with economies that were subordinated to the needs of the industrial nations.[85]

Some of the effects of this colonialism persist today. They can be seen, for example, in some of the pattern of commodity exports from developing nations. The theory of comparative advantage, which underlies the current globalisation of world trade and the various moves to reduce tariffs and other barriers, states that every nation should produce those commodities which it can produce with the greatest relative efficiency. Free trade between these nations will then be of mutual benefit, as the price of goods will be reduced for all participants. In practice, however, some of the historical operations of economic colonialism may have had the effect of destroying certain types of commodity production and re-orienting subjected economies towards other types of commodity production geared to the needs at the time of the colonial power. As the costs of reconstructing the economy around the production of other and possibly preferable commodities may be relatively high for a developing nation, such contingent inefficiencies may persist after decolonialisation.[83]

A related consequence of the depression of the 1870s and 1880s was the increasing competition between the emerging advanced industrial nations. These nations competed for geographical influence and control of resources for later economic development. This process reached another point of crisis with World War One. A short period of stability followed, partly because the combination of war damage and reduced industrial production allowed industrial re-expansion to be fuelled by domestic markets, before the major depression of 1929 to 1932 and World War Two. This was accurately predicted by Marx, although his belief that capitalism would not survive these crises appears to have been incorrect.

Economic theories of environmental damage

All scientific theories have limitations (see Chapter 2). This tends to be especially true of the social sciences, for a number of reasons. The data is usually messy, the observer is invariably enmeshed in the system which is being observed, which makes observations relatively subjective, and it is often impossible to run proper control experiments, which means that data is usually limited and non-random. This has led to a situation where there is often little consensus as to how to explain phenomena, and a number of diverse approaches and disparate theories of social and economic behaviour. These approaches vary, amongst other ways, in the situations and problems they are aimed at explaining or solving, and in the extent to which they are reductionist.

One important debate concerns the degree to which markets should be regulated, which is highly relevant to any discussion of sustainability. Keynesians and monetarists both agree that markets constitute a more efficient way of distributing goods and services around society than command systems, but have a degree of disagreement as to how markets do and should operate. For example, in Keynesian economic theories, governments are seen as having a vital role in market operations. Providing welfare benefits to the unemployed, for example, is a form of legitimate direct government intervention in market operations. Keynes was motivated to develop this concept of the mixed or welfare economy partly by the cycle of economic crises described earlier and by the social and human costs of industrial capitalism.

This approach is rejected by free-market theorists, who argue that any regulation of a market will reduce the efficiency of that market, because the imposition of political or any other kind of superordinate constraints will make it impossible for the market to achieve the optimal matching of resources to demands. This theory suggests that a laissez-faire approach, with unregulated markets, will generate more wealth, which ultimately works in the best interests of everyone. In principle, this additional wealth could then be divided by society in various ways. Libertarian economists argue that if society as a whole becomes wealthier, then, even if there are no redistributive measures, some part of that wealth will eventually 'trickle down' to the poor.

In practice, while it is certainly true that a badly regulated market is very unlikely to perform efficiently, it is also true that an unregulated market probably cannot achieve economic optimality, for reasons that are discussed further later in this chapter.

Some political and economic philosophies have been extended in attempts to include explanations for environmental damage. As the theories are quite diverse, they tend to propose different solutions.

For example, socialist and Marxist analysts have argued that environmental damage arises from corporate pursuit of profit. Marx argued that removal of the parasitical rentier class would allow a more equitable distribution of wealth, thereby enabling a transition from capitalist exploitation of human and natural resources to disinterested scientific management, in the form of state planning, of these resources. In the event, a number of problems arose with the rather flawed and partial application of this model in the USSR. These problems included the emergence of various subgroups such as the *nomenklatura* which acted in their self-interest, the ideological insistence on centralised state planning with the associated inability to respond flexibly to changing circumstances, and the perceived need to meet political targets with the consequent sacrifice of essential flows of information.

Free-market theorists disagree with the socialist analysis. Some free-market theorists have argued that some of the worst environmental damage results from the activities of a state sector that is unresponsive to market forces and thereby to the expressed wishes of the people.

The historical evidence does not entirely support either position. It is clear that environmental damage in countries with command or otherwise centralised and state-directed economies can be as severe as or more severe than in capitalist economies. Some 16 per cent of the geographical area of the former Soviet nations is very severely environmentally damaged.[86] Centralisation of the decision-making process in itself will tend to ensure that the environmental impact of policies is widespread.

However, the history of corporate operations in relatively unregulated environments, such as in developing nations, is not dissimilar. For example, consider Brazil's 'Operation Amazon' project. This was a vast project which was designed to open the Amazon to multinationals, settlers, and ranchers. One component of the scheme, the Carajas iron ore project, was to convert an area of forest the size of France to industrial and agricultural use. Parts of Operation Amazon, notably the land settlement, are now generally agreed to have been both ecologically destructive and economic failures, incapable of generating real profits (due to the low stocking rates achievable) and dependent on continuing government subsidies. Such private sector-led schemes have clearly been partly ideologically motivated, probably as much so as with the state capitalist schemes of the USSR.

In practice, neither the state nor the corporate body is likely to operate in the interests of unrepresented parties, such as the environment or future generations, without some mechanism for representing those interests. If this mechanism does not exist, perceived current demands will tend to outweigh future needs.

Regulated markets

It is likely that if sustainability is achievable there will be more than one way to attain it. Sustainability, in effect, describes a group or class of possible outcomes (see p 38). The best solutions might be those that reconciled the requirements of sustainability with maximum economic efficiency.

There are several reasons for believing that a market-based economy may prove to be more able than a command economy to adapt to the parameters of sustainability and simultaneously achieve a relatively high degree of economic efficiency. For example:

❑ Markets tend to be more efficient than command economies in certain respects. Demand and supply are determined as a result of a large number of individual decisions and interactions which provide more sensitive feedback mechanisms. A command economy is far less precise, being based on a single decision process, which can result in considerable inefficiency and wastage.
❑ The complexity of natural and human ecology is likely, for the foreseeable future, to exceed the capacity of any single institution to develop an adequate model or to exercise effective control. Market economies have a more diversified decision-making process. This greater flexibility may assist in accommodating the new demands of sustainability.

However, there are costs associated with diversification of the decision-making process, especially in the loss of mechanisms of coordination and planning, with consequent possible duplication and other inefficiencies. Markets often arrive at sub-optimal solutions. As discussed earlier, many products are known to be imperfect in engineering or ergonomic terms, but have come to dominate markets and set industry standards through historical contingency (see p 151). A regulator, on the other hand, would be able to impose a change-over, as when, for example, the UK adopted Standard International units of measurement. A further common example occurs when industries converge on whichever of a number of competing solutions happens to gain the greatest market dominance, often without regard to technical or other merits. Examples of this include standards for railway gauges, for example, where the outcome of the commercial competition between railway companies lead to a standardisation on a gauge that was sub-optimal in engineering terms, or video cassettes and personal computers.

It is important to note that such imperfections do not arise out of anti-competitive practices and product suppression, but out of legitimate market competition.

It is unlikely, therefore, that the much more complex demands of sustainability will be achieved simply through the operations of the market.

The optimal solution, achieving maximum economic efficiency within the parameters of sustainability, is therefore likely to be in the form of a corrected or regulated market system. This is further discussed on p 158.

Strategic imperatives for sustainability

The report of the United Nations World Commission on Environment and Development (better known as the *Brundtland Report*)[87] defines sustainable development as development that meets the needs of the present without compromising the ability of future generations to meet their own needs.

This definition, as developed in the report, links together a number of important factors. These include population growth, resource use, environmental impacts and patterns of development. Development issues provide an important focus. These include the nature and role of technology, the distribution of productive potential and of consumption, and the political and economic inequalities between the nations.

The report then develops seven strategic imperatives;

1 To revive growth in the developing nations, in order to alleviate poverty and thus reduce pressure on the environment.
2 To change the definition of growth to include notions of equity and non-materialistic values.
3 To meet essential human needs for food, housing, energy and so on, and to accept that this will entail changing consumption patterns.
4 To address the issue of population growth, especially by changing the economic pressures to have children.
5 To conserve and enhance the resource base.
6 To develop the necessary environmental risk-management technology, and to transfer this technology to the developing nations.
7 To integrate economic and ecological factors in decision-making.

The *Brundtland Report* also lists seven preconditions for the above imperatives;

1 A responsive political decision-making process.
2 An economic system that does not generate the same pattern of resource demand as the current industrial economic system.
3 A responsive social system that will maintain cohesion by redistributing the costs and benefits of unequal development.
4 A system of production that can operate within ecological limits.
5 Innovative developments in technology that permit increasingly energy- and resource-efficient solutions.
6 An international order that maintains cohesion globally.
7 A pattern of responsive, flexible, and self-correcting government.

The *Brundtland Report* represents a considerable advance in thinking, not least because it clearly accepts that achieving sustainability will require a strategy, rather than a piecemeal approach. Given that achieving sustainability will often involve trading-off objectives (see p 189), and that trade-offs can only usefully occur within a strategic context (see Chapter 9), this is a helpful development. However, there are several unresolved problems with the *Brundtland Report*. The most important problem is that there is little real analysis of the structural relationship between the economy and the environment. The report assumes that economic growth can co-occur with or even enhance growth in certain types of natural capital, and that increased consumption in the developing nations (provided that they follow an appropriate development pathway) will therefore be possible without incurring the kind of costs that have been associated with economic growth in the industrial nations.

However, this all depends (as the *Brundtland Report* does suggest) on the kind of economic growth proposed (see p 97).

Organising for sustainability

Given its relative lack of structural analysis, the *Brundtland Report* should not be taken as a blueprint for sustainability. It is, rather, a statement of principles. As a general rule, the purpose of such broad statements is to stimulate progress. It is usually desirable that such statements are eventually superseded by more detailed prescriptions for change. As part of this process of development, it sometimes becomes necessary to return to and revise some of the initial principles.

Given these reservations, the general principles of sustainability outlined in the *Brundtland Report* suggest that certain models of market organisation, with particular roles for government and the private sector, and certain patterns of corporate ownership, ethos and management style are more likely to be conducive to sustainability.

As an example, consider the need to develop new technology for environmental risk management and for increasingly energy- and resource-efficient industrial solutions.

❑ There must be a strategy and a mechanism for directing investment into the development of the necessary technology. This implies a strong role for government in strategic planning, coordination, and market regulation (which may include consumption or credit controls), in order to drive constant innovation in all sectors towards reducing costs and greater efficiency, especially in the use of energy and materials.
❑ Firms must then be able to respond to these signals. A major reconstruction of the industrial base will require long-term investment, which will require appropriate fiscal and other changes. In addition, such long-term investment must be enabled by

addressing the current institutional failures. Firms must be encouraged to develop an appropriate corporate strategy. This will require reducing any pressure for short-term increases in dividends, which in turn may require that the majority of corporate shares are in supportive hands to reduce the possibility of predatory takeover, especially if a high level of long-term investment temporarily depresses share prices. It may also require good inter-organisational relationships, with corporate stake-holders such as employees, customers, investors, bankers, and government representatives routinely informed and consulted, and a corporate ethos that is more collective, has a greater sense of concern for people, society, and the environment, and is more oriented towards planning, experimentation, and development.

Market systems

There are a number of different market systems in the world today, based partly on the different socio-cultural understandings of the relationship between the individual and society. Most of these market systems form parts of mixed economies, where the state plays a role in market regulation but the extent and type of intervention varies. The different market systems have their own combinations of strengths and weaknesses, which can cause them to behave differently under changing circumstances. Some of these market models have more of the features listed above, and may therefore be better able to accommodate the emerging demands of sustainability.

For example, the nations of the G7 group (the major advanced industrial economies, currently Canada, France, Germany, Italy, Japan, the UK, and the US), have different models of market organisation. These can be roughly divided into three groups, as follows;

❑ the relatively deregulated model followed by the 'anglo-economies' of Canada, the UK, and the US;
❑ the corporatist model, followed by Japan;
❑ the social market model, followed to different degrees by Germany and France while Italy contains elements of both this and the first model.

The deregulated economy

The market system currently predominant in the UK has certain key relevant features:

❑ Markets are relatively lightly controlled or are self-regulated.
❑ The interests of the shareholders (in practice, the major shareholders) are normally paramount. Other corporate stake-holders, such as

employees (and, in practice, the minor shareholders), tend to possess few rights and are not usually well-informed or properly consulted, although they clearly have various interests in the performance and behaviour of the company.
❏ The performance of directors is assessed primarily on the basis of dividend and profit performance. This can place directors under considerable pressure to sacrifice training and long-term investment, or to liquidate accumulated capital investments, in order to maintain high levels of dividends.

These features give rise to a system that is strongly financially-oriented and essentially deal-driven.[88]

The corporate economy

The version of capitalism developed in Japan, partly as a result of management ideas imported originally from the US, differs in a number of areas:[89]

❏ The Japanese economy is relatively tightly organised, and markets are managed. The approach was developed for post-war reconstruction, when available resources were directed into successive sectors of the economy on the basis of a national strategy, and continues in a more relaxed form today.
❏ The relationship between corporate stake-holders, such as employees, customers, investors and bankers is long-term. A large firm in Japan will typically have half of its shares in supportive hands, removing the threat of hostile takeovers.
❏ In Japan, the firm is seen as a key part of social organisation. This is unlike the current situation in the UK, for example, where the function of the firm is more accurately defined as being to maximise profits, and in which relationships are expressed in contractual terms that define relative financial reward and the organisational power of the individual.

The Japanese approach, sometimes called peoplism, permits a more constructive attitude to investment. Longer pay-back periods are tolerated, as are lower dividends, in order to achieve corporate growth and a greater share of world markets. A combination of competition between *keiretsu* (corporate trading groups) and financial security within *keiretsu* encourages constant innovation. In the long run, this drives down costs and hence gives increased profit margins, fuelling further growth. The internal market within a *keiretsu* also encourages innovation, by providing a supportive environment for the development of products that can later be launched nationally, and then on to world markets. Some 10 per cent of all sales in Japan are within one of the six large *keiretsu*.

EcONOMIC DEVELOPMENT AND THE ENVIRONMENT

The result can be seen clearly in relative capital investment commitments. In 1980 Japan's capital investment totalled $185 billion, compared with $322 billion in the US. By 1990 the US's capital investment had grown to $544 billion, but Japan's capital investment had grown to $702 billion. This disparity is both absolute and relative. In 1990 Japan invested 25.3 per cent of its GDP, compared with 11.2 per cent for the US and 10 per cent for the UK.

This long-term approach to investment has generated high rates of economic growth. In the mid-1980s the Japanese economy was approximately half the size of that of the US. If the relative growth rates of the 1980s had continued, however, the Japanese economy would have overtaken that of the US in size at some point between the year 2000 and the year 2005.

The social market

The social market model, exemplified by Germany, has certain key characteristics:

❑ The underlying concept of the social market is that all members of that society should in some way benefit from that mode of economic organisation and the flow of wealth that it generates. This makes all members of that society stake-holders, which in turn affects attitudes and actions. Trades unions tend to be consulted to a greater extent than in the deregulated economies, and those that are included in the decision-making process tend to be more reluctant to take industrial action.
❑ Banks are key stake-holders. They have patterns of shareholdings in companies that are not typical of the deregulated economies. Industrial groups often have cross-shareholdings, with banks and bank groups playing a central role in forming these links.
❑ Banks are committed by this type of involvement to long-term industrial and business lending. This changes the pattern of investment. In the UK, for example, there are 4 main clearing banks. Of these, Lloyds has the most long-term loans, with nearly 20 per cent of their loans intending over five years or more. The Midland Bank, in comparison, has less than 5 per cent of its loans over the same period (1988 figures). There is a significant disparity between even this most long-term of the UK clearing banks and the German norm. For example, the largest commercial bank in Germany, the Deutsche Bank, has more than 50 per cent of its loans extending over more than four years.[90]

Economic blocs

One of the major current economic trends is towards increased globalisation of trade. There is another trend, however, possibly equally powerful and potentially in conflict with the first, which is towards regionalisation of trade in economic blocs.[91] Should the latter trend predominate, then it is likely that, in the immediate future, the world will be increasingly economically dominated by the three emerging major blocs. These are the North American Free Trade Association (NAFTA), the east Asian and Pacific nations, and the European Union (EU). Other blocs, such as a Central and South American bloc, may eventually emerge, but this is unlikely to happen immediately.

The three market models currently employed in the G7 group are likely to be influential in the development of these major economic blocs, as each of them is currently dominated economically by one or more of the G7 group nations. The major blocs would then be grouped as follows.

❑ The NAFTA group consists of Canada, Mexico, and the US. Future expansion may include some of the Central or South American states, with Chile currently the most likely candidate for admission to the group. This economic bloc currently appears likely to continue with a relatively deregulated, laissez-faire market system, characterised by, for example, relatively large income disparities (see p 177).

❑ The East Asian and Pacific group is not yet formally organised as a bloc, but has been integrating since the 1970's. In 1991 inter-Asian trade exceeded Asian–US trade for the first time. This group comprises a number of nations, including Japan, Korea, Taiwan, Hong Kong, Singapore, Thailand, Indonesia, and Malaysia. China, Vietnam, and Cambodia are gradually being drawn into this group. The Association of South-East Asian Nations (ASEAN) will probably form part of the nucleus of this emerging bloc. The member nations currently employ various market models, ranging from the Chinese family-based *hongs* to the Japanese *keiretsu*, with the Korean firms in the process of evolution from family-run operations to professionally-run operations. Japan's corporatist, group-oriented approach to economic management will probably exert a strong influence on models of economic development in the region.[92]

❑ The European Union currently comprises Germany, Italy, the UK, France, Spain, the Netherlands, Belgium, Portugal, Greece, Sweden, Austria, Finland, Denmark, Ireland, and Luxembourg. The European Free Trade Association (EFTA), which comprises Switzerland (which includes Liechtenstein in a customs union), Norway, and Iceland, is now partly integrated with the EU. Poland, the Czech and Slovak Republics, Hungary, Bulgaria, Rumania, Latvia, Lithuania, and Estonia (and Yugoslavia before the outbreak of war) are already included in a number of economic agreements, and may eventually join the group in some capacity. The Russian Federation is developing a number of trade links with Europe, and it is possible that in the long term institutional links will develop as well. The European group will therefore eventually comprise somewhere between 19 and 40-odd nations, possibly in a multiple-tier structure that reflects the degree of economic convergence. The need to integrate, harmonise, and regulate the component national economies will create a convergence about a market model. Current indications are that the convergence will be towards some version of a social market approach.

As the UK is a member of the EU but has a market model that more closely resembles that of Canada and the US, it is in a somewhat anomalous position.

Of course, the blocs are dissimilar in a number of respects, including current population, economic size, and economic growth rates. The NAFTA group currently has a total population of 360 million and a gross GDP of $6.2 trillion. The east Asian nations (not including China) have a combined population of 510 million and a gross GDP of $3.7 trillion. The European group (including the current EU and EFTA nations) has a total population of 380 million people and a gross GDP of $6.5 trillion. The current growth-rate disparities are relatively large. If they were to be continued, the east Asian bloc would have a larger economy than the NAFTA group by 2011 and than the greater European group by 2016.[93] Other variables, such as the relative percentages of GDP allocated to military spending, may be significant but are not reviewed here.

The European Union

The model which is likely to increasingly influence the economy of the UK is that of the EU.

The EU states are becoming increasingly economically and politically integrated. Although there have been doubts and political problems with this process, it currently appears likely to continue. The abolition of remaining internal barriers to trade and movement, in the creation of a single European market, has made it necessary to reduce the relative economic disparities between member states. This is because labour and capital would, in the absence of any internal barriers to their free movement, otherwise tend to migrate within the EU away from depressed or less advanced regions towards prosperous or more advanced regions.[66] While the theory of comparative advantage states that regional specialisation will lead to greater wealth for all, as goods would be produced to lowest cost, and thus raise real wage rates throughout, the theory cannot sensibly be applied to situations in which both labour and capital are mobile. Capital mobility in particular is likely to lead to growing wage disparities and other differences between regions, which might undermine political support for the existence of the EU.

This means that the economic harmonisation of the EU member states is likely to remain a political priority, given that one of the most important reasons for the creation of the initial trans-European structures was to reduce the potential for conflict within Europe.

There is currently a 'cohesion' programme, consisting of a series of measures designed to reduce the economic disparities between member states. A number of policy instruments are used to effect this economic harmonisation, including, for example, the Regional Development Funds. These channel funding for regional infrastructure development to those states and regions that are, according to certain indices, relatively disadvantaged.

European integration

The move towards increased integration will have a large range of social, political, economic, and environmental effects. For example, consider the development of integrated environmental standards. In the move towards the single European market, many companies are spreading their operations and production costs across Europe. There is already pressure to establish certain kinds of plant in countries with lower pollution control standards and hence lower compliance costs. For example, some German companies are relocating to southern Europe, due partly to the additional costs of meeting the higher standards of environmental protection in Germany. This damages the interests of both countries concerned (economically in one case, environmentally in the other), and will displace rather than solve environmental problems, so it is likely that there will be further initiatives to counter this trend.

Furthermore, as environmental regulations will become an increasingly significant component of industrial production costs, inequalities in operating costs caused by variations in environmental standards will increasingly be regarded as unfair competition. This is likely to result in commercial pressure to harmonise environmental standards and improve market coherence.

There will clearly raise a number of issues:

❑ No company or nation will be able to commit high levels of expenditure (beyond the current margin of economic viability) to raising environmental standards unless competitors do likewise.
❑ Companies and nations that currently operate with lower standards may resist what they perceive to be an excessive degree or rate of change. This is for a number of reasons. For example, if the imposed level or required rate of investment is too high or too fast, companies may become vulnerable to predatory purchasing of stock as dividends and share prices may be driven down during the investment period.

❏ Nations with high environmental standards may come under pressure to relax these if it appears that they are being used as a barrier to trade. This could have locally unfavourable environmental impacts.

❏ The tightening of EU environmental standards may displace investment. For example, some industrial processes might be relocated into developing nations outwith the EU with limited ability to protect themselves or with over-riding economic need.

❏ One possible mechanism for correcting the displacement of investment would be the application of environmental standards to imports. It is possible that this would be seen as protectionism, and might therefore stimulate retaliatory measures. It would also be necessary to ensure that such a mechanism did not impede trade agreements that were made to encourage and support developing nations (see p 175).

❏ Such environmental tariffs may also infringe protocols of the General Agreement on Tariffs and Trade (GATT). While minimum global environmental standards may eventually be integrated into GATT, it is likely that, for the immediate future, these will be set at relatively low levels. This is because the global disparities in environmental standards are far greater than between member states of the EC. It would be relatively difficult for those countries that currently have low environmental standards to meet high standards, given that these countries also tend to be poorer. This means that high international environmental standards might be a barrier to imports from such countries. The primary purpose of GATT is to promote flows of trade, and such flows of traded goods would be impeded significantly less by environmental standards that were set at a level below those standards currently operated by the economically advanced nations.

❏ The creation of the single European market is likely to stimulate renewed economic growth. This may in turn generate a need for more stringent environmental policies to counteract the extra pollution that will be generated by this increased economic activity. For example, one European Commission report projects an 8 to 9 per cent increase in SO_2 emissions and a 12 to 14 per cent increase in emissions of NO_x by 2010 as existing emission control policies are increasingly overtaken by rises in the demand for electricity and vehicles. This, of course, is a general problem associated with economic growth. For example, if an economy grows at 2 to 3 per cent per annum, and if industrial output grows in proportion, then industrial output will approximately quadruple over 50 years. This means that it would be necessary to reduce the total environmental impact per unit of GDP to approximately one-quarter of the level achieved at the start, over the same period, in order to ensure that the total environmental impact of the economy was no greater at the end of that time than it was at the beginning.

Economic integration and the UK business sector

It was indicated on p 159 that the UK is in a relatively anomalous position within the EU in terms of market organisation. This means that the progressive integration of the UK economy into the EU will require a degree of adjustment, and will expose UK companies to various pressures.

For example, consider the implications of moves to harmonise environmental standards. In a number of EU member states, notably Denmark and Germany, the growth in levels of environmental concern and awareness is increasingly being reflected in a number of legislative and fiscal measures. A number of EU companies have calculated that increasing consumer awareness will create continuing pressure for high environmental standards in consumer goods. Many of these companies are now making environmentally-driven strategic investments, including restructuring operations and the introduction of cleaner technology.

If environmental standards are agreed for the EU, and if such standards are either higher than current UK levels on introduction or are then subsequently raised over current UK levels, then some UK companies may find that they have become uncompetitive.

These pressures would probably be registered in a number of areas:[94]

❏ Inputs to the organisation. These may be affected by absolute shortages of resources (should some resources move towards exhaustion), imposed restrictions on resources (through legal measures, for example, to limit the use of certain products), and implied restrictions (where certain patterns of resource use become socially unacceptable, and change is imposed through public opinion with changing customs and ethical standards).
❏ Organisational processes. The move to cleaner technology, for example, might be encouraged by changes in the pricing or taxation structure.
❏ Outputs from the organisation, in the form of organisational products and services. This is where the main source of change is likely to be in the immediate future. Possible factors include rising costs of waste disposal, increasing difficulty in getting some wastes dealt with at all, rising emission standards, the possible introduction of costly pollution licences or permits, bans on certain dumping routes, and tougher environmental protection for certain habitats.
❏ Total inputs to and outputs from the organisation. Manufacturing firms and other organisations process flows for which they do not currently account (in fact, most organisations are not aware of many of those flows, especially those that are uncosted). A baker, for example, processes flour into bread in a recognised flow, but also processes water, energy, chemicals and metals (in depreciating machinery), and produces waste heat and emissions of dust in partly recognised or unrecognised flows. These unrecognised flows are likely to become more salient with rising environmental standards and costs.

❏ There will also be implications for business strategy, taxation, pricing, and investment appraisal.

Many UK companies still see environmental issues as being peripheral to them and their concerns. Any significant change in the UK's position in this regard, such as a commitment to meet rising EU environmental standards, would mean that those firms and other organisations that are not adequately prepared would tend to find that the best recruits go elsewhere, sourcing and supply would become increasingly problematic, and that the markets for some products and services would shrink. Of course, even the most unresponsive firms eventually respond or cease to exist.

The UK business sector would be well advised, therefore, to adopt one of the currently available environmental management system guidelines. These currently take the form of voluntary codes of conduct, but there is likely to be increased social and consumer pressure to adopt such standards, and to move towards greater transparency and improved environmental reporting. In the longer term, it is probable that standards for environmental auditing being determined by the European Commission will eventually be further tightened in the form of mandatory directives.

All changes of this kind, of course, represent both threats and development opportunities. The advantage to a firm in carrying out environmental audits and adopting environmental management system guidelines lies partly in identifying potential problems, and partly in identifying future development opportunities. In general, firms that do take such pro-active steps must enhance their prospects of surviving the period of change, establishing a lead in new fields of activity, gaining public relations benefits, and benefiting from a higher level of staff awareness, involvement, and commitment.

12

Socio-cultural Factors

In some cases there are relatively simple and obvious connections between actions and effects. With many environmental issues, however, action and effect may be widely separated in space or time. For example, decisions made in the UK today may have effects over the next century, or in another part of the planet. This introduces an important element of uncertainty. In such cases, it may be considered appropriate to act without what would normally be considered to be adequate evidence, as it may be impossible to obtain such evidence. It could also mean that those affected, such as a future generation, a less-developed nation or a relatively powerless group in society, would have little or no means of retribution or redress. Such factors therefore require guidelines for behaviour that are not purely objective or utilitarian. If a given generation accepts certain constraints on its use of resources, or if a powerful nation accepts certain responsibilities to some much weaker nation (to ameliorate some environmental impact, for example), then it is likely that these will in part be guided by some principle of responsibility. This means that it is important to consider such cultural factors as ethics and values.

This chapter reviews these issues, considers some of the mechanisms whereby ethical positions could be translated into policy, and identifies some of the redistributive consequences of extending these principles domestically, internationally, and intergenerationally.

Ethics and equity

The issues reviewed in Chapters 4 and 5 are likely to give rise to a number of difficult decisions; each will have a different distributional effect. Every possible decision will cause a redistribution of costs and benefits in space (geographically) and in time (across generational boundaries). This raises a number of questions, concerning ethics, value judgements and cultural understanding of the relationship between the individual and society.

Intergenerational issues are particularly important in the question of sustainability, because they contain elements of all of these questions. The current deforested state of the UK, for example, is the result of historical economic operations in which past generations appropriated existing natural assets and transferred the associated costs (environmental damage, opportunity costs and so on) to the current generation, arguably without adequate compensation. Similarly, current economic operations routinely improve profitability by appropriating natural assets and by transferring the associated costs on to future generations.

This temporal inequity could actually be reversed in a transition to sustainability. The reason for this is as follows. It is likely that a transition to sustainability will require some investment in new technologies and industries. Obviously, this programme would only be undertaken if it were perceived to be likely to lead to general benefit at some future time. However, there are certain possible scenarios in which the interests of the future generation may conflict with those of the current. For example, it may be that new understanding of environmental limits and some impending ecological disaster indicates a need for a rapid pace of investment. This in turn may necessitate a short-term reduction in average consumption and utility. Given that the generation that is first to commit itself to achieving sustainability effectively assumes the burden of environmental debt inherited from previous generations as well as accepting responsibility for its own behaviour, there is likely to be a degree of inequity at whatever point the decision is taken.

This scenario has important implications. It means that society would have to decide how the reduction in consumption should be distributed both domestically and internationally. Decisions would have to be made about who should make what kind of sacrifice. This in turn introduces an ethical question as to the extent to which the current generation are prepared to accept responsibility for those who would be disadvantaged by particular developments, for future generations, and possibly for other forms of life.

Games theory

Games theory analyses the formal characteristics of rules of behaviour. In essence, by defining the rules of behaviour in social and economic systems, for example, and the motivation of the participants in some systems, it becomes possible to predict outcomes.[95]

One games theory model is known as the 'prisoner's dilemma'. This models a situation in which there are intense benefits and diffuse costs. This represents elements of a conflict between the interests of the individual and those of society, and, by extension, between those of current and future generations.

Consider the following example:

❑ I am an individual car owner. I derive very significant personal benefit from my car, but must suffer some of the collective consequences of mass car ownership, including congestion, acid deposition, lead pollution and so on.

❏ Everybody would be better off collectively if everybody gave up their cars, as there would be no congestion and no pollution, but I personally would be much better off if everybody except me gave up their cars. There would be no congestion and minimal pollution, and I would still have all of the personal advantages of car ownership.

❏ All car owners are now asked to voluntarily give up their cars, for the good of the community and the environment.

There are now four possible outcomes if everyone except me acts in the same way. These, ranked from most desirable to least desirable, from my point of view as an individual car owner, are as follows:

1 Everyone except me gives up cars. I would get both the collective and the personal benefits.
2 Everyone gives up cars. At least I would get the collective benefits.
3 No-one gives up cars. No change. I don't benefit, but I don't lose.
4 No-one except me gives up cars. I get no collective benefit, and I also suffer the personal cost.

Each individual car owner now has to decide whether to 'cooperate' (trust that everyone else will give up cars at the same time) or 'defect' (decide to keep cars). In such a situation, the way most people reason, in effect, can be summarised as follows:

❏ My decision is independent of everyone else's decision.
❏ If everyone else defects, then I am better off defecting than cooperating, because option 3 is better for me than option 4. Similarly, if everyone else cooperates I am still better off defecting than cooperating because option 1 is more advantageous for me than option 2.
❏ Therefore I will defect and keep my car.

This line of reasoning is essentially rational, given this particular set of rules. Of course, this rationality at the level of the individual can lead to sub-optimal, illogical and undesirable outcomes at group, societal, or planetary level.

Repeated studies of the 'prisoner's dilemma' have shown that it is usually very difficult to persuade people to cooperate rather than defect.[96] In general, trust must be built up, which is sometimes possible in iterated prisoner's dilemmas, when participants interact repeatedly; otherwise some other mechanism must be found to ensure cooperation. In practice, the social pressure to maintain such cooperation can only be applied when an overwhelming majority of the population is committed to cooperation. For example, if parks are generally litter-free, it is often possible to persuade people who would otherwise litter not to do so. It is also usually possible to tolerate a low level of defection within such a hypothetical community. However, as soon the numbers defecting increase significantly, and the park starts to fill with litter, the social pressure diminishes and a widespread outbreak of defection and littering tends to follow.

This analysis suggests that exhortation and education may not be sufficient to persuade enough people to change their behaviour and to sacrifice their perceived personal benefits for a communal good. If this is so, intervention may be necessary, in order to provide incentives for cooperative behaviour and disincentives to discourage defection.

Alternatively, some external motivating factor may be introduced. If the participants in a prisoner's dilemma game are prepared to make sacrifices or take risks, then cooperation can be secured. This might be obtained if, for example, there were some factor (such as the welfare of their children or a moral or religious code of ethics) that was sufficiently powerful to persuade participants to act in other than their immediate perceived self-interest.

Selfishness, altruism, and sustainability

The question that arises from earlier in the chapter is whether the current generation is likely to undertake the initial commitment to a transition to sustainability. Related questions include the extent to which current members of society have chosen to maximise consumption at the possible expense of future generations, and why the current generation employ high discount rates, which give reduced consideration to the effects of decisions on future generations.

There are a number of relevant factors:

❑ Understanding of the subject is currently poor. This could affect the position in various ways. For example, perceptions of the seriousness of the global environmental crisis might be significantly affected by perceived time horizons. Events which are believed likely to occur within the projected lifespan of the individual decision-maker may be accorded a higher priority. Simple lack of knowledge is itself likely to result in a relatively low priority being accorded to environmental issues.

❑ There is also a possibility that there may be a common misunderstanding of the nature of the relationship between humans and the natural environment. Human perceptions are deeply conditioned by social and economic systems. If society has had little real cause, to date, to take full account of environmental issues, then people might be expected to accord a low priority to sustainability. However, there would be no reason why this situation should necessarily be regarded as natural or immutable.

❑ Some societies have operated with a zero or negative discount rate. For example, many of the sacrifices made by the Russian people in the years after the revolution that created the communist state were made in the belief, at least in some cases, that they would create a better standard of living for future generations. To sacrifice personal utility for the sake of children or grandchildren represents the application of a negative discount rate to decisions. The planting of parkland with hardwoods, where the aesthetic benefit will not mature for a number of generations, similarly requires the application of a negative discount rate (see p 125 and p 129).

❑ People will not necessarily have a single set of preferences that apply to all decisions. It is possible for one set of selfish preferences to apply in the market place, and for another set of socially responsible or altruistic preferences to apply in public life. This was a model followed, for example, by some of the industrial philanthropists of the Victorian era. Some people will vote for a political party that has policy goals that would restrict their personally favoured options, that is, for a government that would oblige or enable them to behave more ethically. This may be based on the reasonably rational conclusion that personal sacrifice may be noble, but less effective than group sacrifice at achieving results. Group sacrifice, additionally, may preserve the relative position of individuals within that society. There are various possible explanations for the existence of these multiple preferences.[97, 98] One theory is that there is a dissociation in human thought between private, economic decisions, and public, social decisions. Another is that the majority of non-pathological human beings allow themselves to maximise their self-interest within the bounds of agreed socially responsible and ethical behaviour. A third is that most decision-making is only partly rational, and is influenced at least as much by uncalculated emotional commitments and unexamined assumptions. Certainly, the calculation of costs and benefits of abstract economic rationality bears little resemblance to real human psychology (see p 180).

Implementing equity

Assuming that society made a commitment to implement some form of intergenerational equity, there would be a set of complex decisions to be made as to how this equity could be achieved. For example, one possibility is that current expectations of rising consumption be replaced by expectations of constant consumption. This does not of course mean that current levels of consumption are themselves necessarily sustainable, just that stabilising at a given level is more likely to be sustainable than indefinite increases.

Constant consumption might then be achieved by investing (rather than consuming) the rents made from the depletion of non-renewable resources into those forms of artificial capital which can substitute for these resource inputs in the production function. There are two problems with this approach. One is that it does not take into account any loss of environmental amenity that might be associated with the substitution of artificial for natural capital. The other is that there may be limits to the extent to which resources can be substituted.

Another possible approach, therefore, is to set a condition that some measure of the stock of natural capital must remain constant. This does not suffer from the same technical problems as the first approach, but does not yet provide a sufficient answer, as trade-offs between the various forms of capital are likely to be unavoidable in practice (see p 103).

A more fundamental complication arises with the need to develop an index of consumption (or other index of welfare) that could be compared across generations. This could be either a mean or a median measure of per capita consumption, for example, depending on whether it is decided that it is only aggregate consumption that matters, or whether one should have regard to the distribution of that consumption. It would be possible, for example, for aggregate welfare to increase (if society as a whole became more wealthy) while the distribution of that welfare became less equal (if the affluent members of society became much richer, while the condition of the poor members of society improved only slightly, remained the same, or even deteriorated).

There is a logical inconsistency with a position that accepts a responsibility for intergenerational equity, but none for intragenerational equity, as that would imply that the current generation were (uniquely) worth less than possible future generations, which would in turn reveal a lack of intergenerational equity (unless one adopts a narrower definition of intergenerational responsibility, as being confined strictly to one's direct descendants).

This implies that if society is to accept responsibility for intergenerational equity, it must also have some regard to domestic and international equity and distribution of resources.

There are certain situations, usually in developing nations, where the connections between intergenerational and intragenerational equity are relatively clear. For example, it is known that very poor people may be driven to over-stretch their environment. Deforestation for fuel wood, over grazing of marginal pastures, and overpopulation are all associated with poverty and lack of access to resources, technology, information, and so on. It is also often the case that the poor are the victims of environmental pollution. As the poor tend to be politically powerless, they are rarely in a position to challenge the effective ownership of the environment by those wealthy people who control commerce and industry. If environmental property rights were to be transferred to the poor, or held in a common trust, then charges could be levied on those who used the environment's pollution absorption capacity. This would result in a flow of revenue to the poor, and would create an incentive for sources to reduce their emissions.

In these cases, redistributive measures may be effective. If resource property rights were to be re-allocated to the poor then they would have a source of income, and a degree of control over and a reason to invest in their environment. In these relatively simple cases, the same action is likely to alleviate poverty, reduce disparities in wealth, and improve the environment, all of which would probably help society to move towards a more sustainable position. In complex, post-industrial economies, however, it will generally be more difficult to identify a single course of action that would have a similar multiple effect. Thus moves towards intergenerational and intragenerational equity in an advanced economy will require careful calculation of the likely costs, consequences and benefits of each step.

A further extension of the idea of intragenerational equity is that it should be applied at a global level. It is unlikely that the planet could generate the resource flow or has the pollution absorption capacity to allow every human to enjoy the standard of living of the industrial nations. Intragenerational equity therefore would require a reduction in current consumption and pollution emission levels in the wealthy nations. There are various ways in which this could be done.

One possible method for reducing emissions from the developed nations, for example, would be to establish a set of global pollution permits, which could be traded internationally. There would be a number of important political and technical issues involved, however, in the choice of the mechanism for implementing a more equitable distribution. Take emissions of carbon, for example. These can be calculated on a number of bases, including aggregate, per capita, or per unit of gross national product (GNP). The choice of basis is particularly important. In terms of aggregate totals, nations such as India and China contribute nearly as much to the greenhouse effect as the US and the former Soviet bloc. However, on a per capita basis, the average citizen of the US produces four to five tonnes of carbon each year, considerably more than the average citizen of India (0.4 tonnes) or China (0.6 tonnes). The reason for the large aggregate contribution from India and China is that these two nations combined contain more than one-third of the human population of the planet.

Similarly, Canada and Australia are currently ranked thirteenth and twentieth respectively of the worst offending nations in terms of aggregate emissions, but second and third respectively in terms of per capita emissions. On a per capita basis (and also on a cumulative basis), the greater part of global carbon emissions are from the industrial and OPEC nations.

If a global carbon emission quota were to be set, for example, at 0.5 tonnes of carbon per person per year, the USA would be operating at a large margin over quota, whereas India would be in credit.

If carbon emission permits were traded internationally, nations that were over quota would have the choice of either investing in the necessary control technology or purchasing additional carbon emission permits from countries such as India, which would have unused quota to sell.

Depending, of course, on the price attached to pollution permits, this measure could reduce disparities of wealth between nations, generate a flow of funds to nations which generally find it hard to secure the capital necessary for development, provide an incentive to reduce pollutive emissions, and thereby improve the environment. All of these measures (see p 157) would probably help a general transition to a more sustainable and equitable distibution of rights and resources.

At a cash value of $15 per tonne, and a quota of 0.5 tonnes per person, the USA would have to invest in emission reduction or buy in quota to the value of some $13–17 billion each year, while India would receive more than $1 billion each year from retailed quota.

Some might object that this would provide an incentive for the less-developed countries to increase their population – or, at least, a disincentive for such nations to curb their population growth. This would probably, in practice, depend on the marginal value of each additional human being when the increased income from the allocation and subsequent retail of carbon quota was deducted from the cost of that human being. This is only likely to be a factor in the very poorest countries, such as Mozambique, for example, which has an average per capita annual gross domestic product (GDP) of $78 (1988 figures).[99]

However, if emissions were to be calculated on a per unit of GNP basis, the flow of funds would be largely reversed. India, for example, generates more than twice as much carbon per unit of GNP as the US. Japan would be one of the major recipients of funds under such a scheme. This is partly because the less advanced industrial nations tend to have more obsolete and less efficient manufacturing plant, and partly because some of the more polluting manufacturing operations have been transferred into or recently established in developing nations where labour is cheaper and environmental standards are lower.

The concept of tradable pollution permits and the various distributional effects of different applications are further explored on p 118.

Equity and technology transfer

The issues discussed on pages 168 and 172 suggest certain potential conflicts of interest. The idea that efforts should be concentrated on a search for technological solutions to environmental limits, for example, is more likely to seem attractive and feasible to a developed nation than to an underdeveloped nation.

This is because the developed nations tend to be at least partly dependent on continuing flows of natural capital from the less developed nations, which represent costs to the consuming nations, which would therefore be set off against the costs of developing the technology. A developed nation is also more likely to have depleted its own resource base, which would therefore no longer represent an alternative to imports. Finally, a developed nation is more likely to have the technological infrastructure to make such developments possible.

Some of the underdeveloped nations, however, have a very different perspective. Some of them see the current prominence accorded to ecological issues as a new form of economic imperialism, in that it looks like an attempt to deny to them the development pathway already pursued by the economically powerful developed nations. Similarly, the attempt by the developed nations to develop technological substitutes can look like an attempt by the natural capital consuming nations to extend their control over commodity prices by reducing demand. Any reduction in demand is particularly serious for a national economy that is heavily dependent on a small number of exported commodities, a situation that may well have resulted from the last era of economic imperialism. Furthermore, some of the undeveloped nations have less current and historical responsibility for global environmental problems, and perhaps therefore feel less of a moral onus to act. Finally, cost implications for underdeveloped nations tend to be quite different from those for developed nations, as an underdeveloped nation is less likely to have the technological infrastructure necessary to make such developments possible.

Such considerations, amongst others, have prompted several developing nations (India has been prominent amongst these) to request technology transfers from the developed nations to permit them to achieve the global environmental goals for which they have accepted an appropriate level of responsibility. These requests have met with a very limited success. This is because the developing nations have lower labour costs than the developed nations. A transfer of state-of-the-art technology would therefore allow the developing nations to supply their own domestic markets, thereby reducing the market available for exports from the developed nations, and, eventually, to become effective competitors with the developed nations for a bigger share of the world market. These technology transfers therefore pose a sharp conflict of interest for the developed nations.

Such conflicts of interest could be resolved in an appropriate context. For example, the carbon emission quota system reviewed on p 172 would require those nations that exceeded their quota (which, if the quota were calculated on a per capita basis, would mostly be the industrial and OPEC nations) to either invest in pollution reduction or to pay for the excess, at some appropriate rate. It would, of course, then be possible for the funds that accrued under such a system to be simply disbursed to those nations that did not use their full carbon quota (which, on a per capita basis, would mostly be the underdeveloped and the developing nations), and for these funds to be at least partly used to fund purchases of advanced technology. It would also be possible, however, for payments due under such a system to be made in kind. For example, direct technology transfers to nations that performed below carbon quota could be credited at an appropriate value against the sum due. This could allow nations that were over-quota to meet their obligations at less real cost, as some of the technologies concerned would be of higher value to the recipient nation than to the donor nation.

Such contexts do not currently exist. However, it may be necessary to develop such funding and compensatory mechanisms in future, given that the most likely alternative to a global 'cohesion' programme of this type is increasing competition for diminishing resources and for a greater share of the planet's pollution absorption capacity, and an unequal allocation of social, economic, and environmental costs, with the consequent potential for tension and conflict.

Welfare

The fundamental questions raised by the concept of sustainability, and especially by the discussion of values, make it important to consider the purpose of human economic activity. In a relatively traditional society the purpose of human economic activity is usually fairly obvious: to allow exchanges of both essential and socially-important goods and services, preferably with a sufficient margin above subsistence level to allow the enjoyment of some leisure and recreation.

In a complex diversified economy the purpose of many exchanges is divorced from subsistence or other purposes of relatively unambiguous or intrinsic value. For example, the greater part of international world trade by value occurs in manufactured goods exchanged between countries who both have the capacity to manufacture these goods, such as cars from Japan to the US and Europe. These goods are shipped between these countries for several reasons, including the following:

❏ marginal price differences, which exist because low energy costs keep transport costs below that margin.
❏ fashion and styling considerations, induced wants, and perceived value and quality.

Much of this economic activity is not addressed to meeting essential human physiological and psychological needs. This is not the same as saying that it does not contribute to human welfare. However, much of this economic activity probably does not contribute to welfare either.

For example, more money spent on medicine does not necessarily correlate with the health level of the population. Recent fashions for expensive and increasingly esoteric vitamin and mineral supplements are not thought by the medical profession to be likely to lead to better health or higher intelligence. Similarly, more money spent on education does not necessarily mean that society will automatically become more skilled and educated. At least some of the money going into education in a market-based educational system is likely to be defensive expenditure, that is, expenditure judged necessary to preserve the relative positions of individuals. Expenditures in these two areas taken as examples could therefore give some indication of both possible increases in welfare and possible decreases in welfare.

In those areas where wants must be focused on styling and fashion considerations and induced by advertising, expenditures are more likely to indicate levels of psychological need for status and reassurance, rather than real welfare. Psychological research, together with the fact that levels of expenditure in these areas continue at high levels, indicates that the emotional reassurance derived from a particular pattern or act of material consumption is transient. This is probably because psychological and emotional needs can only be adequately addressed and satisfied at a psychological and emotional level. The displacement of psychological and emotional needs into patterns of material consumption will undoubtedly generate considerable economic activity, and a consequent energy and resource cost, but cannot properly satisfy these needs.

The current definition of economic welfare, which essentially measures absolute consumption, therefore overestimates 'real' welfare (see p 95) by including items of negative as well as positive utility, and by including material substitutes as satisfiers for emotional needs.

Income and wealth

The section above raised some of the issues with the assessment and measurement of welfare. Most conventional economic models take income and wealth to be good indices of welfare in a market-based economy. While these are arguably not fully adequate as measures, they clearly measure an important component of welfare. In general, the aggregate sum of human goods and services is taken to represent the total potential economic welfare, while the distribution of this aggregate sum determines the welfare of individuals.

As discussed on p 172, it would be possible for the aggregate sum of goods and services (the total potential economic welfare) to increase while the distribution of welfare became narrower and the welfare of some people therefore diminished. This could happen if, for example, society as a whole became richer but the wealth became increasingly concentrated, leaving the poor poorer.

For a number of practical reasons, the question of the measurement of welfare and that of the distribution of welfare (and the mechanisms for the distribution of welfare) must be considered in conjunction.[100] For example, in severely underdeveloped regions or countries, it is important to include measures of consumption (rather than just measures of income) in indices of welfare. This is because increased savings often do not, in situations of great poverty, feed forward directly into increased development potential. In practice, they tend to be absorbed into consumption until a perceived surplus is generated.

In many cases this perceived surplus is likely to be achieved more rapidly by institutional changes that increase equality than by rises in production, if the additional surplus (and hence income) generated by such rises in production is unequally distributed in the direction of greater inequality. For example, the World Bank's World Development Report 1991 states that 'there is no evidence that saving is positively related to income inequality or that income inequality leads to higher growth. If anything, it seems that inequality is associated with slower growth.' Inequalities in incomes also tend to underestimate the extent of inequalities in welfare. For example, there is a strong relationship between levels of income and wealth, and health. Low incomes correlate with poor health. There is no consensus as to why this relationship obtains, but in general poor people are more likely to eat poor diets, live in poor-quality housing, have less access to health care facilities, and so on. A further possibility is that low status and consequent low self-esteem and high stress levels themselves undermine health, a possibility that is borne out by recent work in psychoneuro-immunology.[101] Excessive inequalities in income can therefore generate increased costs, in the form of indefinite remedial programmes, increased health care demands, lost economic potential and so on.

In spite of this, various theories of economics and of economic behaviour have held large disparities in wealth and income to be necessary as a means of motivating performance and to reward those with scarce skills. Large disparities have therefore been justified as a means of improving economic performance. There are related arguments to the effect that the primary motivation for economic performance is absolute levels of wealth, rather than relative levels of wealth. From this it would follow that economic dynamism would lead to greater inequality.

This theory, translated into policy, has contributed to a number of measures that have had the effect of increasing inequality in the UK. Between 1981 and 1989 the ratio between the earnings of the top decile and the earnings of the bottom decile of the UK workforce grew from 3–1 to 4–1, one of the largest increases in disparities in Europe.[102]

The psychological literature, however, indicates that people at a sufficient margin above subsistence levels of income are more influenced by relative rather than absolute disparities of reward. For example, senior levels in a number of professions are rewarded at rates set by reference to the rest of the profession. At very high salary levels it is probably the case that the status associated with a particular salary level is the important motivating factor.

It is also clear that many people are prepared to work for lower levels of income if they have a greater sense of identification with the group, and if their contribution is recognised in other than income terms. For example, Japanese income grades tend to be flatter than in the UK. The ratio of incomes between the top two deciles of the workforce and the bottom two deciles of the workforce is about 2.6–1 in Japan, compared to some 4.4–1 in the UK (1980 figures). This lower income disparity is possible partly because the Japanese corporate system provides acknowledgement and recognition in other than material terms, and places greater emphasis on the role of the team rather than the individual. A comparison of the post-war economic performance of the UK and Japan indicates that high income disparities do not necessarily correlate very highly with economic dynamism.

In conclusion:

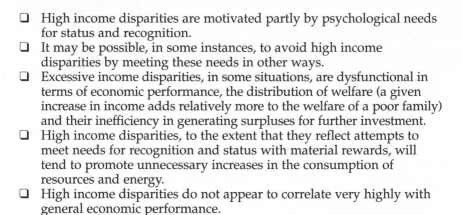

❑ High income disparities are motivated partly by psychological needs for status and recognition.
❑ It may be possible, in some instances, to avoid high income disparities by meeting these needs in other ways.
❑ Excessive income disparities, in some situations, are dysfunctional in terms of economic performance, the distribution of welfare (a given increase in income adds relatively more to the welfare of a poor family) and their inefficiency in generating surpluses for further investment.
❑ High income disparities, to the extent that they reflect attempts to meet needs for recognition and status with material rewards, will tend to promote unnecessary increases in the consumption of resources and energy.
❑ High income disparities do not appear to correlate very highly with general economic performance.

This suggests that transition to a more equitable distribution of income, wealth and resources, both domestically and internationally, could be commensurate with – and possibly even promote – more investment and more dynamic and efficient economic performance.

Limits to neoclassical economics

Philosophers have fought and wrangled,
An' meikle Greek an' Latin mangled,
Till, wi' their logic-jargon tir'd,
And in the depths of science mir'd,
To common sense they now appeal,
What wives and wabsters see and feel

Robert Burns, on a debate between
Adam Smith and Thomas Reid, in a
letter to James Tennant of Glenconner.[103]

Neoclassical economics, which is the core of current thinking in economics, is a particular philosophical and methodological approach that provides a number of useful tools for understanding some of the issues reviewed in this book. Chapter 6 examined some of the ways in which a standard neoclassical approach could be developed and extended to provide tools for dealing with environmental issues, and thereby assist in a transition to a more sustainable way of life.

There are, however, a number of problems with the neoclassical approach, and some ways in which it may not be entirely adequate to the task. Some of the more specific of these issues were discussed in Chapter 6, but there are other, more general problems with the neoclassical approach in the context of sustainability, which are reviewed in this section.

The essential problem is that while a neoclassical analysis provides a powerful range of tools and methods for understanding and hence changing economic systems, it can only be extended to wider analysis of social systems by making a set of controversial assumptions. This may make it necessary to import other theories of human behaviour from other disciplines, and either partially reconstruct the current principles of neoclassical economics or develop a model in which economic information can be combined with environmental, social, and psychological information.

The assumptions that underpin a strict neoclassical approach, and that may limit the extent to which it can be extended into the social, political and environmental dimensions of the sustainability debate, are as follows:

❑ Most neoclassical economists assume that human preferences are formed exogenously (outwith the domain of economics), as if they appeared at birth. This means that preferences are not supposed to be themselves influenced or formed by economic systems. This approach, therefore, does not allow for the possibility that fundamental preferences might be formed or shaped by social, economic, and cultural factors such as education or advertising, possibly in a recursive manner (so that preferences might be shaped by economic systems, and in turn go on to shape economic systems, and so on). The assumption that preferences are formed exogenously is very important, because it is these preferences that define utility and social welfare.

The presupposition that human preferences are formed outwith economics, a subject of enquiry that then concerns itself with the expression of human preferences, is based on the Newtonian–Cartesian concepts of pure and abstract reductionism and objectivity. These concepts were introduced widely into the social sciences during the early years of this century, largely because of their success as metamethodological models in the natural sciences. However, these definitions of pure reductionism and objectivity are no longer generally shared by natural or other social sciences. This means that strict neoclassical economics employs concepts that no longer form part of a common methodology. An approach that would be more compatible with other social science methodologies is to allow that preferences may be formed or shaped by social, economic, and cultural factors, which brings preferences into the same dynamic matrix of events.

❑ Many neoclassical economists assume that the forms of economic functions (such as utility and production functions) do not change over time. This is a similar assumption to the one above, in that because human preferences are assumed to be exogenous, as if they were fully-formed at birth, there is no reason to suppose that they will vary between cultures or over time. From this it would follow that one could assume that future preferences would probably be the same as today's preferences. This assumption, too, is not generally shared with other social sciences. If preferences do vary between cultures, and over time, then it would follow that the form of economic functions could also vary.

❑ Many neoclassical economists assume that individuals derive happiness from absolute, rather than relative levels of consumption. Research in psychology, sociology and social anthropology, however, indicates that change relative to others in society, and relative to expectations, is usually more important to people than change in absolute levels of consumption (above those levels necessary for survival).

❑ Scientific enquiries usually observe the boundaries of disciplines. In general, therefore, many economists tend not to inquire too far into the complexity of human psychology. As a result, many economists make no distinction between the kind of trade-offs or substitutions that may be possible on psychological or ethical grounds. Some trade-offs, for example, may be inequitable because humans often estimate risk and probability inaccurately, while others may be unethical, because they ignore the legitimacy or otherwise of various forms of ownership. There is an ethical argument, for example, that the natural resource base cannot be considered to be legitimately and exclusively owned by the current generation in the way that the artefacts produced today could be considered to be owned by the current generation. This point is relevant to the discussion of time and utility discounting on p 125. There is a similar question as to whether it would be more ethical to accord other species an existence value, in addition to any other defined value that they might have to the human species. The deep ecology ethic, for example, holds that all forms of life have an inherent and inalienable right to existence. Other arguments suggest that the concept of intrinsic value should extend to include non-biological features of nature as well as biological features.[104] Any extension of values beyond human interest would clearly pose serious problems for an approach that is based on a comparison of relative utility for one species.

❑ Much of the neoclassical approach in economics is based on the philosophy of utilitarianism, and is underpinned by a theory of human motivation called hedonistic associational psychology. Over the last 50 years utilitarian philosophy and hedonistic associational psychology have become less accepted and less widely used in other social sciences. This means that neoclassical economics has concepts of welfare, optimality, and rationality that, in common with the assumptions as to the nature of human preferences and the form of economic functions, no longer form part of a common philosophical and intellectual base.

❑ Many neoclassical economists argue that analysis must be limited to variables that can be quantified. Unfortunately, this can mean in practice that precision is obtained by excluding large numbers of factors. This may become increasingly significant, given the current limits to human knowledge of the potentially important dimensions of sustainability. This, of course, is a common problem in most disciplines. What is particularly relevant in this case is that there are a number of factors, such as social stability, which are not immediately amenable to quantification and so are more likely to be excluded from the analysis. It is exactly these factors, however, that may be very important for human society and which may therefore determine the success or failure of programmes of transition towards sustainability.

❑ An analysis of complex problems, such as social or environmental problems, that reduces the variance to one dimension (such as money, energy or utility) entails a particularly high cost in information. There is an especial risk that this procedure can have the effect of reducing real uncertainty to apparent certainty by excluding relevant information, thereby limiting decision-making attention and excluding actual conflicts of emotions, values and ideologies. An analysis of monetary impacts is usually relevant, but it may not be safe to assume that this analysis will satisfactorily capture every dimension of a complex problem that may include changes in the nature of society and changes in non-monetary resources, such as natural capital stocks and flows.[105]

❑ Similarly, there are problems with an approach that assumes that it is possible to compensate those who must bear the cost of pollution control measures or the opportunity costs of depleted natural capital out of increments of value accruing to others. This approach depends on the assumption that 'correct' prices or values exist, or that politically-determined values may be used. However, as has been discussed in earlier sections, markets are not perfect for a number of reasons. These include the existence of monopolies and oligopolies, the perpetuation of existing inequalities, institutional or historical barriers to innovation or mobility and the influence of existing economic geography. All of these factors will have the effect of distorting prices. This means that 'correct' prices may not occur in real exchanges, and that politically-determined values may be influenced by existing inequalities.

The consequences of environmental damage and the benefits of environmental controls may also be heterogeneous, that is, of quite different natures. If consequences and benefits are of different types, it certainly makes quantitative comparisons difficult. It may, under some circumstances, make such comparisons invidious.

There are a number of factors, such as psychological well-being and aesthetic values, and social features and functions such as community cohesion, integrity and identity, which are of different natures to those functions for which quantification and compensation is generally seen as appropriate. Such factors, once compromised, may be irreplaceable or irreparable and so beyond compensation.[105, 106] If such factors are to be protected, this implies imposing certain restrictions on the use of the conventional cost–benefit analysis. Cost–benefit analysis will clearly be appropriate in some circumstances, but ultimately requires that there is a consensus in society (and especially amongst the entire group of those affected by a given decision) about the evaluation rules inherent in cost–benefit analysis.

Unfortunately, this consensus cannot be obtained across future generational boundaries. This means that it is difficult to justify the use of cost–benefit analysis on issues which have significant long-term effects that run across generational boundaries.

13

A Systems Approach to Managing Sustainability

Systems theory and decision-making

Formal systems analysis, as with scientific methodology generally, currently plays a relatively minor part in political and economic decision-making. In many resource and pollution management issues, for example, a typical assumption is that the systems affected will respond in linear rather than non-linear fashion. Awareness of the implications of long-term gradual change is often imperfect, and system elements are often treated discretely, instead of as parts of a system. This tendency, as has been discussed, is particularly problematic when dealing with non-linear systems.

Some of the more significant problems are associated with the general understanding of probability and causality. With probability, for example, the stated degree of risk is often accorded more significance than other important factors, such as the number of times that the risk is incurred. This can lead to risks not being properly accumulated and understood.

Similar problems obtain with causality. The innate structuring function of human perceptual and cognitive systems predisposes us to see meaning and causality in proximate events: there is a certain tendency to attribute causal connections to events that occur in sequence. Scientific rigour and objectivity to some extent controls for this tendency, but it is a significant factor in much human thinking.

Complex systems can generate outcomes that depend on numerous interactions, and which may therefore be highly sensitive to the precise starting conditions and loading of factors. Certain classes of complex systems will, in addition, behave chaotically under some conditions. This means that models of many types of complex system, and chaotic systems in particular, are probabilistic rather than precisely specified, and limited in certain respects. With weather systems, for example, forecasts are probabilistic and cannot be projected indefinitely into the future without rapid loss of precision.

The tendency to attribute simple causality can prevent the development of an understanding of probability and interactive causality in complex systems.

However, a number of attempts have been made to introduce elements of systems thinking into organisational management. These fall, roughly, into two groups – the hard systems approach and the soft systems approach.

Hard systems

The hard systems approach is essentially about defining the problem-solving sequence.[108, 109] This involves some version of the following steps:

1 *Problem definition.* This involves defining the problem and what has to be done.
2 *Choice of objectives.* This involves deciding what would be required to reach each objective, and formulating the measures of effectiveness which then form the basis for making comparisons between strategies.
3 *Systems synthesis.* This involves identifying the various possible alternative systems.
4 *Systems analysis.* This involves analysing and evaluating the various hypothetical systems in the light of the objectives.
5 *Systems selection.* This means choosing the most promising alternative.
6 *System development.* This entails developing the chosen alternative up to the prototype stage.
7 *Current engineering.* This consists of the realisation of the system, but also includes the essential processes of monitoring the system, feeding this information back to the design stage, and then modifying the system itself as necessary.

This basic model can be extended in various ways, to include, for example, indices derived from welfare economics, or other multiple criteria.[110] Much political and economic decision-making is done on the basis of multiple objectives, some more consciously-realised than others. However, such decisions are often made with very little knowledge as to how these objectives can and do sometimes conflict. As a general principle, every decision implies one or more trade-offs. It would represent a great improvement over much current practice if the inescapable trade-offs between objectives happened in a conscious and deliberate political or management decision-making process.

The role of this kind of systems analysis is not to replace the decision-making process, but to provide decision support; that is to improve the quality and range of information available and to make the decision-making process more transparent and robust.

A slightly different version of this approach, developed by the RAND corporation, focuses on the costs and benefits of alternative programs. It involves the following steps:[111]

1   *Defining objectives.* This involves clarifying the desired aims and goals.
2   *Describing the various alternative techniques or systems* available for achieving those objectives.
3   *Identifying the costs and resources* required to achieve each alternative.
4   *Developing system models*, in the form of a mathematical or logical framework that can show the interdependence of the objectives, the systems, the environment, and available resources.
5   *Developing the criterion for selection*, and relating the objectives, the costs, and the resources to choose the optimal or otherwise preferred alternative.

A number of problems arise when these hard systems approaches are applied to soft systems, especially those systems that involve humans. The hard systems approach starts with a basic acceptance of the objectives, problem specification, and organisational needs. Hard systems engineering aims to provide a solution to a defined problem in the terms in which the problem is posed, so these factors are generally taken as given. With soft systems, however, there are frequent disagreements as to what the goals and objectives should be. It is very important to recognise this issue, and deal with it, so that the whole process does not degenerate into what Merton calls 'the quest for improved means to carelessly examined ends'.[112]

Soft systems

With soft system applications, system thinking should be regarded as a contribution to problem-solving, rather than as a goal-directed methodology.

This applies to all situations where the task itself cannot be entirely and objectively defined. Where a hard systems approach can be used to deal with structured problems (for example, how to get this piano up this flight of stairs), a soft systems approach is necessary when dealing with unstructured problems (for example, how to define the aesthetics of this piece of piano music). With unstructured problems, the definition of the problem and the designation of the objectives is itself problematic. Where the definition of the problem depends on the viewpoint adopted, it is important to make that viewpoint explicit, and to then work out the systemic consequences from that point.

This also means, in practice, that the precise sequence of stages in the analysis of the situation and in the development of solutions cannot be maintained. This is because the problem itself usually gets redefined during the process, so it is necessary to be prepared to go back to the first stage, and spend more time reconsidering the problem. It may even be that the client for whom the problem is being solved gets redefined as part of this process. Systems in which humans are involved are always multi-valued. Formal organisations, for example, exist for a stated purpose, that is, to carry out a defined function. However, all such organisations have additional reasons for existence. No organisation consists solely of formal associations between people made in order to effect some particular task. Organisations also comprise important human relationships. Part of the purpose of many human interactions is to establish and modify relationships, which usually happens in conjunction with any formal and acknowledged goal-directed purpose. It is common, for example, for organisations to resist structural change, as such change threatens existing power structures and thereby conflicts with the informal personal agendas, and to accept only procedural changes that do not threaten the maintenance of a particular distribution of power and status within the structure. This means, in practice, that change must be both systemically desirable and culturally feasible.

The stages of a soft system approach, then, are more like the following:

1   Reviewing the unstructured problem situation.
2   Clarifying and expressing the problem situation.
3   Defining the relevant systems and subsystems, whether these are formal or informal.
4   Building conceptual models, scenarios and analogies.
5   Comparing these models with the expressed situation.
6   Effecting such changes as are currently both feasible and desirable.
7   Taking action to improve the problem situation.

This process involves defining and redefining the objectives by building models, developing criteria, comparing the models with the current situation, and both reconstructing the models and altering the current situation in a process that involves continuous backtracking, iteration, and feedback.

In summary, soft systems methodology has more feedback than a hard systems approach, with continuous comparisons being made between stages. This kind of systems thinking is itself an input into organisational change.

Soft systems are more general than hard systems. A hard systems analysis can be used when the problem is highly defined. For example, a hard analysis of how to get a piano up the stairs will always produce a way of getting the piano up the stairs (assuming that this is actually possible). A soft analysis could redefine the problem as one of getting the musician and the piano together, and arrive at the solution of asking the musician to come downstairs.

This means, in practice, that an open-ended approach is needed, where the the outcome is not seen as being an optimal solution to a particular problem, but a continuous learning process.

Any soft system model, fundamentally, will always be defined in terms of a given Weltanschauung, a world view that provides a context within which events are given meaning. The flows through the human elements of the system are not neutral information, but events with complex clusters of meaning for the participants. A given task or operation might be precisely defined at one level, but may still have quite different meanings for the human actors and participants. That element of the definition of a function or task that depends on one's perception is termed the *latent function*. For example, where there are three bricklayers engaged in the manifest task of laying bricks, the latent function for one might consist of earning so much per hour, while the latent function for the second might consist of laying bricks, and for the third of building a cathedral.[113] This applies to system structures too. Most human structures are, ultimately, embodiments of beliefs and perceptions. Prisons, for example, are embodiments of concepts of crime and punishment.

Checkland, addressing this need to define human systems in human terms, has identified six elements that should be explicitly described as part of what he terms the root definition:[15]

1   *Transformation of inputs to outputs*. This involves identifying the main flows into, through and out of the system.
2   *Ownership of the system*. This involves identifying the decision-makers and stake-holders.
3   *Actors in the system*. This involves identifying the wider community involved with or influencing the system.
4   *Customers of the system*. This involves identifying the demands that people make of the system, for later comparison with the stated purpose of the system.
5   *Environmental constraints on the system*. This entails identifying all constraints in the social, economic and natural environment of the system.
6   *Weltanschauung*. This involves discussing and clarifying the world-views, perspectives and perceptions of the participants, owners, actors and customers.

In summary, therefore:

❑   The investigation of the social world is in some ways fundamentally different from that of the natural world.
❑   The classic analytical reductionist approach is often inappropriate for dealing with problematiques, which include elements of both natural and social systems.
❑   In social systems, such factors as roles, norms, values, concepts and applications of power are all critical. It is better to address this issue than to try to hide it in the decision-making process. This often makes it necessary to use more participative planning techniques.
❑   As a general rule, systems cannot be defined until the system they serve has been defined. This has a number of implications for human and management systems. For example, when designing a planning or an information flow system, it is best to start by considering the organisation to be served by the planning or information system, before going on to consider what information is required, from where, in what form, how frequently, and so on. This means that it is important to define the functions and roles of the humans in the system before designing the service systems.

Information aggregation and decision-making

The task of achieving sustainability can be separated into two key stages, as follows:

☐ The identification of policy options that will lead towards greater sustainability.
☐ The development of appropriate mechanisms, strategies, and techniques for implementing such policy options.

The first will require the development of a better understanding of the behaviour of complex natural and human systems and the various interactions between such systems. This will require a great deal of further research. Existing theories will probably have to be extended to encompass the additional information that this research will generate. It may also require new techniques, methodologies, and theories.

The second will require a number of political developments. It will also require the development of techniques and processes to introduce the information generated by a better understanding of system behaviour into actual decision-making processes.

There are currently a number of ways to model human social and economic systems. All existing techniques, given the current deficiencies in our knowledge of the behaviour of interacting complex systems, may prove to be inadequate. This, of course, is true of all science. However, in this instance, the interdisciplinary and complex nature of the problem makes it especially important to keep the methodology under review.

Neoclassical economic theory provides a number of tools for modelling some aspects of human behaviour, and consequent pattern of resource demand. Other economic theories, and other social and natural sciences, provide both complementary and alternative tools and methodologies.

The various approaches to environmental evaluation and the related decision-making models may be ranked in terms of the extent to which they aggregate information.

☐ Highly aggregated. This includes conventional cost–benefit analysis, for example, which reduces all variables to one dimension (usually in terms of money equivalents).
☐ Intermediate. This includes various analyses of cost-effectiveness, in such forms as ALARA (as low as reasonably achievable) and BATNEEC (best available technology not entailing excessive cost). This approach bridges non-monetary and monetary values, but does not resolve the basic problem of environmental valuation.
☐ Highly disaggregated. This includes environmental impact statements, or data from Geographical Information Systems (GISs) or other non-monetary information systems, which can be used as separate inputs to decision-making in addition to financial statements. This then permits some form of trade-off analysis, such as positional analysis (see below).

It is important to note that the choice is not between reductionist and non-reductionist models. All models are reductionist to some degree, as information loss must be accepted in order to gain simplicity and clarity. A decision must be made, however, as to how much information can be sacrificed while still retaining an adequate sample of reality. Highly aggregated models have a higher cost in terms of information loss.

The method reviewed in Chapter 14 uses relatively disaggregated information. It is based on the assumption that environmental, political, social, economic, and psychological problems are often interrelated, embedded in each other, or composite (see p 190). This places the emphasis on establishing distinctions between relevant and irrelevant conditions, rather than assuming that these are determined by disciplinary boundaries.

There are a number of decision-making techniques that would be compatible with this approach. The decision-making model reviewed in detail below is called *positional analysis*. It is a tool for analysis and graphical presentation of information from several dimensions (which could be, for example, environmental, social and economic). It must be emphasised that positional analysis is not the only possible decision-making aid, nor the only technique that would be compatible with a general systems approach. It was selected for more detailed appraisal here because of its relative simplicity and immediacy, and the way in which it allows complex information to be presented in a way that is relatively accessible. Given that many of the relevant decisions in any transition to sustainability may have to be taken in the near future, and by people with a variety of skills and backgrounds, these qualities are important.

Composite problems

Composite problems are characterised by:

❑ *Interdependence of factors.* The important factors in composite problems tend to be related directly or indirectly to each other. An attempt to solve such a problem by dealing with only one factor can cause unintended and sometimes undesirable consequences elsewhere in the same system. It is necessary to deal with more than one factor at the same time, therefore, and to approach such problems with a coherent policy framework to coordinate the necessary actions.
❑ *Varying coefficients of interrelations.* Some interrelations between factors will be highly responsive, others will be responsive within certain bounds, while others will be non-responsive. The time factor will vary too. Some factors will respond immediately to change in others, while some factors have more inertia or a weaker linkage and will only respond with a delay.
❑ *Positive feedback and cumulative effects.* These have the effect of causing effects to increment beyond proportion to the original impulse.

Improvements in nutrition amongst very poor people, for example, give rise to increased labour productivity, which lead to increased agricultural and other forms of production, which give rise to improved nutrition, and so on. Such positive feedback usually operates until reaching external limits, or until the system concerned is saturated. Improved nutrition, for example, could lead to an increase in population, which could eventually exceed even the enhanced food supply, which could reduce available per capita levels of nutrition.

❑ *Negative feedback.* This happens when change in one condition is curtailed by change in a dependent condition. To continue the example above, population growth could give rise to increased demand for food, which would give rise to increased pressure on the food source. Should this increased pressure go past the point that the source can withstand on an indefinite basis, the source may then become over-stretched or exhausted. This will give rise to a decline in the food supply, which will give rise to a decline in the population.

❑ *Circular causation.* This is where changes in one condition cause change in other conditions and vice versa in a reciprocating evolving dynamic.

❑ *Non-equilibrium.* Every composite problem is essentially a maladaptive or otherwise undesirable system outcome. All such systems have degrees of dynamic stability, but are never static. Such systems are in a process of continuous change and adaptation. Any attempt to interfere with such a system and to induce a different outcome is, in effect, an attempt to change the system's adaptive pathway by altering significant control conditions in either the internal structure or the external environment of the system. Interventions and consequent changes in such a system can be classed as being endogenous, when they derive from the internal dynamic, or exogenous, when they result from an intervention from outside.

These characteristics mean that the coefficients of interrelations are often not known with precision, and that it is often impossible to be completely certain what the results of some particular intervention will be.

Although the complexity of such situations often means that not all the information is available or quantifiable, this does not mean that the results are limited to the qualitative. Understanding depends ultimately on the ability to measure conditions and changes of conditions, although this is clearly more difficult with an open systems model than with a closed systems model.

Environmental and development issues are good examples of composite problems, for the following reasons:[114]

❑ Environmental and development problems are typically multidimensional, multidisciplinary and multisectoral.
❑ They usually have both non-monetary and monetary dimensions. Many of the non-monetary resources at stake are unique, and the degradation of many non-monetary resources is often irreversible or

difficult to reverse, partly because non-monetary resource depletion often entails an increase in entropy.
❏ They often have spatial and temporal ramifications, extending across politico–geographical boundaries and through time.
❏ They often extend beyond the immediate actors. They may be caused by actors outwith the location of the environmental damage, or may impinge on the interests (such as property rights) of other parties.
❏ They frequently involve conflicts between interests and ideologies.
❏ They generally involve uncertainty and risk, and both the uncertainty and the risk tend to be multisectoral.

Positional analysis

The purpose of any decision-making process is, after taking the relevant facts into account, to arrive at a clear decision that gives the individual or the organisation the best chance of achieving the chosen goals.

The essential function of positional analysis is to illuminate and improve this decision-making process (see Chapter 14). There are a number of methods that have the opposite effect, in that they are generally perceived to be ways of 'solving' problems. This permits a transfer of responsibility from the individual or group charged with making the decision to the method chosen. This in turn allows the choices that are actually being made to be 'lost' in the process of making the decision, as the method chosen will be based on a number of important assumptions that have the effect of excluding certain factors.

Positional analysis makes these choices more explicit. It is a way of representing information and clarifying issues and choices. It is related to a number of broadly similar techniques, such as rapid rural appraisal, participatory learning methods, rapid assessment procedures, and participatory action research. Its specific functions are as follows:

❏ To produce reliable knowledge through systematic enquiry.
❏ To illuminate the decision-making process.
❏ To identify all the options and trade-offs available to decision-makers.
❏ To make the decision-making process more transparent, by, for example, identifying winners and losers, commonalities and conflicts of interest and ethical and ideological issues.
❏ To use complete option profiles on the key relevant dimensions, and to match with the value profiles of the decision-makers in a gestalt or pattern-matching approach.

Essentially, a positional analysis consists of the following steps:

1 Identifying the relevant conditions, factors and dimensions when dealing with a compound problem or making a strategic decision.
2 Establishing procedures for quantifying, measuring, ranking or otherwise prioritising change on each of these dimensions. This can

be done using more than one scale or numeraire. Each dimension should have an appropriate scale.

3   Measuring or rating the options or scenarios in terms of the change on each dimension.
4   Identifying trade-offs on each of the relevant dimensions.
5   Making a decision on the basis of the overall profile of each option, which may involve assigning weights to each dimension. Examples are given in Chapter 14.

Positional analysis makes it possible to use different value profiles (which could be, for example, 'economic growth at all cost', or 'environmental protection regardless of cost') as the targets, and then to monitor systematically progress towards this matrix.

The collapse of a wide range of relevant factors into a single index (such as cash, or welfare) may conceal important features, or accord an effective low value to elements that are less verifiable or less amenable to quantification. The use of multiple indices and accounting procedures explicitly acknowledges the diversity of factors in complex issues, and thereby makes the decision-making process more transparent.

The use of multiple indices makes it necessary to assign explicit weights to the indices used. This may appear to be a more complex process than those methods that use a single index. It is important to note that any single index method must make exactly the same choices. A factor that is excluded (on the grounds that it is difficult to quantify precisely, for example) is, in effect, being accorded zero weight. The difference is that, with a single index method, these choices tend to be determined by the method itself. This makes these choices relatively inaccessible, and less amenable to being questioned.

Another way of looking at this is that the use of a single index means that the decision is taken one stage later than with a multiple indices method, that is, after the indices chosen have been weighted and the scores collapsed into a single figure. The use of multiple indices moves the decision-making part of the process one stage forwards, that is, before the information is 'lost' by being collapsed into a single figure.

The gains in terms of clear and transparent decision-making are valid, irrespective of whether the decision to be made is between two clear alternatives or between a number of options. However, the gains in terms of improved understanding and acknowledgement of choices and trade-offs are likely to be more apparent in those cases where a choice must be made between a number of options, which is more typical of strategic issues.

Information alone, of course, is not always sufficient. There also has to be a will and a mechanism for incorporating this information into real decision making processes and attendant actions. In particular, those organisations that have a 'restrictive' mode of operation tend to be poor at assimilating information from outwith the defined categories in which they see themselves as operating. Much of the information needed to manage a transition to a more sustainable position will clearly be new, and therefore outwith defined categories for many organisations. The application of positional analysis is therefore likely to be insufficient unless accompanied, at least in some cases, by a degree of organisational transformation.

In general, however, while it is true that information is needed to manage change, it is also important to note that the provision of information can itself be an effective way of changing behaviour. This is for two reasons, both of which depend on the relationship between the individual and the group.

First, people often attempt to incorporate new information into their actions, usually going through progressive stages of assimilation and adaption. Second, certain kinds of information, presented in an effective manner, can create an 'information inductance' effect. This means that the information that an individual or group is required to report will influence the behaviour of that individual or group as the actors seek to produce actions which, when recorded by the information systems, will appear benign. For example, corporate managers undergoing divisional performance evaluation tend to behave differently when reporting return on investment figures than when reporting residual income figures. Similarly, some people will exaggerate their business expenses when reporting profit and loss figures to the Inland Revenue, then minimise or disguise them when attempting to borrow money from the bank.

This means that if systems for managing information or making decisions are to be effective at enabling the group to adapt effectively to internal and external needs and circumstances, they must be understood and supported within the group, introduced as part of a coherent strategy for the group and be accorded a high organisational priority.

14

Assessing Sustainability

Sustainability Assessment Maps

The outputs from a positional analysis can be presented in a graphical form. This makes the information more accessible. Sustainability Assessment Maps (SAMs) are a graphic tool for displaying positional information and assisting in decision making. In more detail, their purpose is as follows:

❑ To help to identify the range of effects entailed by each decision.
❑ To clarify the trade-offs that are implicit in each strategic decision.
❑ To assist organisations to define their goals and priorities.
❑ To make all parts of a decision making process clear and explicit. Preconceptions, assumptions, and value judgements are inevitable, but are not usually made explicit. This allows them to influence the decision process in an unacknowledged, unverifiable, and uncorrectable manner. Making all inputs explicit, even without other change, will tend to improve management practice.
❑ To serve as an educational tool.

Essentially, a SAM consists of a diagram in which each of the important dimensions in a compound problem is represented by an axis. Measurements of change or indications of priorities are then mapped onto these axes. The resultant profile can be used to represent the current situation. Possible future scenarios or outcomes from the situation are then used to generate further profiles. These profiles are then differentiated, which highlights the trade-offs inherent in each possible choice.

The first step in this process is to identify the important dimensions involved. These can then be ranked. If some of the important dimensions are environmental, for example, it would be sensible to distinguish between those that measured change in some stock of critical natural capital (where it would be a priority to avoid or minimise any negative change at all) from those that measured change in some non-critical resource (which could be negotiated or traded-off against some gain elsewhere). Similarly, if some of the important dimensions are economic, it would be sensible to distinguish between those that represented factors essential to the continued survival of the sector or company concerned from those that could be conceded against some reciprocal concession.

Information sources can be used separately or aggregated. To calculate all of the direct and indirect, upstream and downstream environmental costs, resource flow and environmental impact implications of a project, for example, would be an extremely lengthy, complex, and expensive process. Many of the methods that would be necessary to do so are not as yet agreed. It will often be necessary, therefore, to sacrifice some information for usability, and to use simplified aggregated profiles to represent complex phenomena. The whole point is to use a standardised procedure for doing so across all the project options, so that all are compared on a like basis. The value of the exercise lies in using axes that allow reliable comparisons to be made between options, and to make these choices explicit, so that it is always possible to identify and check assumptions and calculations.

It is perhaps worth noting that competent management is not the same as democracy.[72] There may be cases where it is legitimate to contain information within the organisation, for example, but to exclude information from the decision-making process itself (by making assumptions, massaging data, or selecting axes of measurement that give the best profile) is self-deluding, and therefore not in the best interests of the organisation itself.

SAMs can also be used to help to clarify points of disagreement between organisations. If all the parties in a dispute about the nature, location, type, and impact of a particular development utilise the same set of axes, then separately identify their preferred positions on each axis, a differentiation of the sets of final patterns will clearly identify areas and degrees of disagreement.

This would allow an organisation, for example, to identify a particular position on a particular axis as being 'inviolable', other positions on other axes as negotiable within limits, and a third group of axes as being of relatively little or no concern.

In addition, SAMs can be plotted with concentric overlays, so that (for example) local, regional, and global effects could be illustrated on a single diagram. This can assist in revealing areas of impact and concern outwith the immediate organisational remit.

Comparisons between the SAMs for different options, or between an ideal SAM and an actual SAM are done on an axis by axis basis, so that the trade-offs become clear.

It is important to note that where a yes or no decision is to be made, the information must ultimately be collapsed into a single dimension. At that stage in the process, SAMs have no advantage over cost–benefit analyses. The difference between SAMs and cost–benefit analyses is in their function as decision-making aids. The purpose of SAMs is to make the trade-offs more explicit, to allow the incorporation of dimensions with which there would otherwise be valuation problems into a single model (by allowing non-equivalent scales), to make the decision-making process more accessible, to assist in the clarification of assumptions and positions, to help in the identification of a full range of options, and to enable the more effective monitoring of the wider effects of decisions over time.

Choosing axes

Various examples of SAMs are given later in this chapter. Not every axis illustrated in the examples will be relevant to every type of project. Equally, there are axes that are not illustrated here that would be relevant to other types of project. What is important is not the exact selection of axes illustrated. Any selection of axes will be at least partly arbitrary. What is important is to make a consistent selection, so that all project options are being compared on the same basis, and to make a selection that reveals, rather than conceals. It would be pointless to exclude an axis on the grounds that it gave a preferred option a relatively unfavourable rating. The purpose of the exercise is to make trade-offs visible, and thereby to make assumptions and subconscious decisions part of the explicit decision-making process. It is entirely legitimate to decide that a relatively unfavourable profile on one axis is outweighed by a favourable profile on other axes, but it is poor management practice to exclude possibly relevant and necessary information before the decision is made.

It will be necessary to decide how to score each project option, how to locate this score on a particular axis, and how the axis should be scaled. In practice, there will be wide disparities between axes in the amount of information available, in whether any consensus exists as to how to take measurements, in the fineness or coarseness of grades available, and so on. It is very important to note that, unlike an analysis that reduces all variance to a single dimension, the axes are not equivalent, nor are the scales necessarily comparable between axes.

Scaling

Axis scales fall into three types, ordinal, interval and ratio. There is a fourth type of scale, the nominal scale, which simply measures identity without ranking (sorting people by religion, for example), but these cannot be used here as they do not reveal priorities.

Ordinal scales

Ordinal scales measure identity and rank. The teaching ability of university lecturers, for example, might be graded as poor, average or good. Ordinal scales are relative, not absolute. To perform well relative to the rest of the field does not give any information as to whether the standard itself is adequate or not. In many cases, however, especially where there is a lack of consensus as to what would constitute an adequate standard, ordinal scales may well be the only ones available in the interim.

Thus an axis could be used to represent a scale of values held by the organisation. Agencies concerned with activities that determine land use, for example, might wish to define a grading system for land quality, perhaps reflecting the degree of statutory protection afforded to certain types of land, as one of the relevant dimensions. This might take a form along the following lines:

1   Site of Special Scientific Interest;
2   National Scenic Area;
3   Environmentally Sensitive Area;
4   Area of Great Landscape Value;
5   other designated or protected area;
6   green belt;
7   green field;
8   existing semi-developed;
9   existing industrial;
10 brown field (previously developed, now cleared);
11 derelict;
12 derelict and contaminated.

Ordinal axes could also be used to represent more subtle organisational change, such as meeting legal environmental standards and requirements, or the introduction of environmental accounting techniques.[88] This would allow an organisation, in two separate exercises, to identify a target or ideal position, and to identify the current position. It would then be possible to differentiate the two profiles and identify areas and extents of any disparities. The reason why such an axis can still be treated as an ordinal scale rather than a nominal scale is that each step can be rated in terms of the degree of difficulty involved and the extent of the transition required. The points on such a scale might look like this:

1   carry out a financial environmental audit, and cost any actual and potential liabilities;
2   establish an environmental budget;
3   introduce an internal environmental transfer pricing system;
4   establish an environmental investment rate;
5   introduce comprehensive external environmental reporting;
6   insist on environmental audits of takeover targets, then suppliers, and even customers;
7   introduce environmental capital asset maintenance accounting;
8   introduce resource flow or input–output analysis;
9   reappraise the product and activity range.

These are, of course, very coarse gradations, and finer scaling would be possible.

Interval scales

Interval scales provide numbers that reflect the difference between scores. This is because the measurement units are equal. Intelligence tests are a good example of this type, as a scale with equal units but no absolute zero. Zero IQ exists in the abstract, but not in reality. One can say that someone with an IQ of 200 is 100 units more intelligent than someone with an IQ of 100, but not that they are twice as intelligent. Similarly, a reduction from 60°C to 30°C in the temperature of a cooling water outflow (when multiplied by flow rate) gives a measurement of a reduction in energy loss. It does not mean that the temperature of the outflow has been halved. Interval scales have zero points, but these are arbitrary. The centigrade scale is arbitrarily set to have 0°C at the freezing point and 100°C at the boiling point of water. Absolute temperature is measured on the Kelvin scale, which is zero at –273°C. A decrease in temperature from 60°C to 30°C is actually a decrease from 333 to 303 K, about 10 per cent (this is not particularly important in this example, as the relevant information here is the temperature relative to the outside).

Ratio scales

Ratio scales are essentially interval scales that have an absolute zero. With a ratio scale, numbers reflect the real ratios between items. To produce two tonnes of waste, for example, is to produce exactly twice as much as one tonne of waste. Examples of changes that could be measured on a ratio scale include the following:

- ❑ profits and losses;
- ❑ tonnages of waste;
- ❑ number of jobs created;
- ❑ hectares of land required;
- ❑ number of deer culled.

Generating options: emissions and environmental impacts

The examples given here are for illustrative purposes. They are based on a number of assumptions, and should not be taken to represent actual profiles.

Ideally, every development or investment option would have an associated comprehensive environmental impact analysis of the direct and indirect effects. This analysis would include, for example, the kind of wastes that would be produced, the toxicity of the wastes (in terms of the lethality and the carcinogenic and teratogenic potential), the volumes of waste, and the time over which waste would be produced. From this would be derived mean and median dilution factors, which would be cross-matched with the disposal route (ground, air or water) to calculate the ecological and biological risk factors and impacts. These would then be subject to site sensitivity analysis and risk analysis (in terms of the most likely outcome and the worst-case scenario, with the associated probabilities). This exercise should be run for local and non-local, short-term and long-term effects. Related analyses should be run on the upstream and downstream consequences of major investment decisions. In practice, much of this will clearly be impractical or impossible. However, it should normally be possible to make some estimated or nominal allowance for effects outwith the immediate situation.

As an example, imagine that a power company is about to install new generating capacity, and wishes to compare three options:

1   A coal-burning power station.
2   A nuclear power station.
3   A tidal barrage.

The first option, the coal-burning power station, will produce large volumes of chemical waste, such as CO_2 and SO_x and alkaline ash. The CO_2 and SO_x will be vented into the atmosphere, in large volumes over a long period of time at a fairly high degree of dilution (using tall chimneys), the ash will also be produced in large volumes over long periods, and will be disposed of in land lagoons.

It is calculated that the local and short-term effects of the airborne pollution will be within the tolerance limits of the local ecology, given that the station is to be sited in a low-grade area which has been used for industry for some time, and that enough of the pollution is going to be exported off-site. The land-based lagoons are a more immediate problem, as the area used will be rendered very alkaline and little plant life will survive. On balance, it is calculated that the risk associated with pursuing this development is relatively low, and likely to remain so.

However, the non-local and long-term effects analysis proves less favourable, as the company accepts at that point partial responsibility for global warming and acid rain damage. The associated risk is relatively high, and estimates are rising as more information becomes available.

This option also incurs considerable additional costs upstream (such as the mining and transport of the coal), each part of which should have an associated environmental impact analysis.

The second option, the nuclear station, has a different profile. The emissions will be complex, including radioactive isotopes and other chemicals. There will be some atmospheric disposal, but the bulk of the high-level waste will be temporarily stored on-site then go to a central underground depository.

It is calculated that the local and short-term effects of the airborne pollution will be below the tolerance limits of the local ecology and below the limits of biological safety for the site. This is particularly important, as the station is to be sited on a remote and environmentally-sensitive part of the coast.

However, the company is less sanguine about the next stage of the process, long-term storage of high-level waste. The risk of disaster is assessed as low, but the worst case scenario is bad. This gives a less favourable risk profile. In addition, the waste product is highly toxic, although the volume of the waste product is relatively low.

Non-local effects analysis then obliges the company to consider the associated upstream costs, such as uranium mining and fuel processing and transport, each of which has an environmental impact.

The third option is the tidal barrage. The wastes produced are associated mainly with the construction phase, unlike the other two options. Considerations of waste toxicity and volume are therefore largely inapplicable.

However, the company calculates that the barrage will flood and destroy the current ecology of some mudflats in the estuary. These mudflats are extremely important breeding grounds for a rare species of bird, and the site of the last known colony of a species of endangered plants. This gives a profile of high site sensitivity and high local ecological damage.

However, the non-local and long-term profile is good, especially when it is noted that the tidal barrage option will only damage one estuary, whereas the fossil fuel option will contribute towards global warming and consequent rise in sea level, which will eventually affect every estuary in the world.

The decision must also take into account the construction and operating costs. Barrages tend to be expensive to build and cheap to operate. Nuclear power was intended to have a similar profile, but safety factors and decommissioning problems have forced up costs further than was originally envisaged. The Severn barrage, for example, would cost some £8.5 billion to construct, against perhaps £2 billion for a nuclear station. However, where the nuclear station would be written off in 20 to 30 years, the barrage would operate over some 120 years. In addition, barrage fuel costs are zero, while nuclear fuel costs are likely to increase over time. Finally, the Severn barrage, when operating, would represent some 8 or 12 GW installed capacity, compared to a 1 GW capacity of a nuclear reactor, although this output would not be continuously available.

The power company develops these profiles and submits a planning application. The planning authority adds a profile itself, in order to compare the option of building any generating capacity at all with the option of investing in insulation and energy efficiency in the homes, offices, and factories in the area concerned.

Axes

For the purposes of this example, the axes chosen are as follows:

- ❑ critical natural change;
- ❑ other natural capital change;
- ❑ site value;
- ❑ aesthetic impact;
- ❑ impact scale;
- ❑ impact risk;
- ❑ emissions (air);
- ❑ emissions (land);
- ❑ emissions (water);
- ❑ emissions (electromagnetic and ionising radiation);
- ❑ emissions (auditory);
- ❑ net capital growth;
- ❑ application of capital;
- ❑ capital commitment;
- ❑ employment impact;
- ❑ total material input;
- ❑ total energy input;
- ❑ resource depletion (fossil fuel);
- ❑ resource depletion (mineral);
- ❑ resource depletion (soils);
- ❑ resource depletion (water);
- ❑ resource depletion (biological).

The preferred outcome (not necessarily the smallest number) is located nearer the centre in each case, and the object is therefore to minimise the score on each axis (not necessarily to minimise the total area covered).

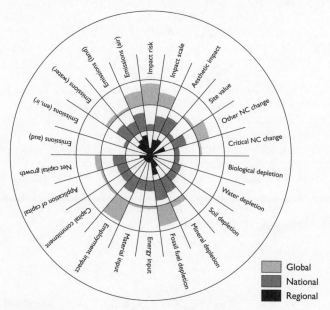

Figure 10: SAM for a new coal-fired power station.

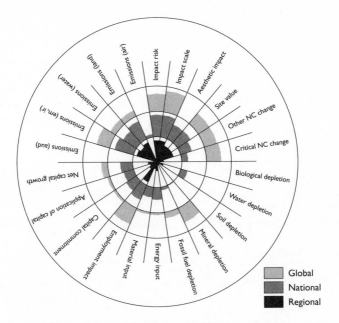

Figure 11: SAM for a new nuclear power station.

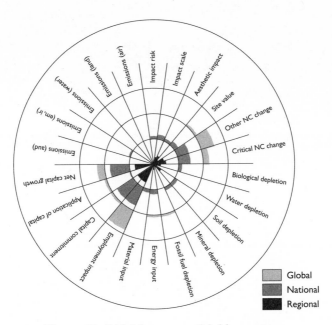

Figure 12: SAM for a new tidal barrage.

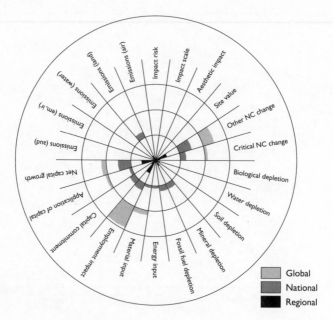

Figure 13: SAM for a major energy conservation scheme.

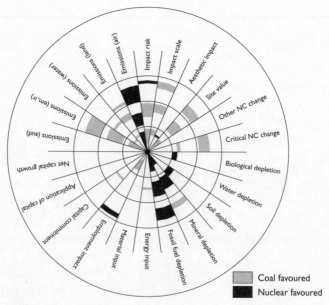

Figure 14: The comparison between a nuclear and a coal-fired power station.

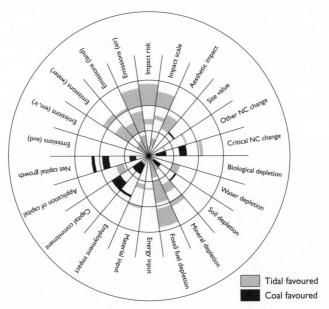

Figure 15: The comparison between a coal-fired power station and a tidal barrage

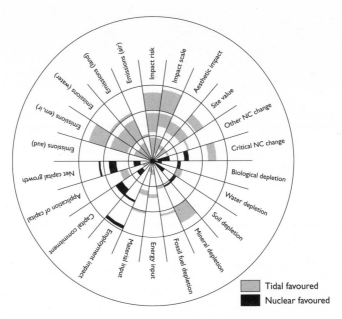

Figure 16: The comparison between a nuclear power station and a tidal barrage

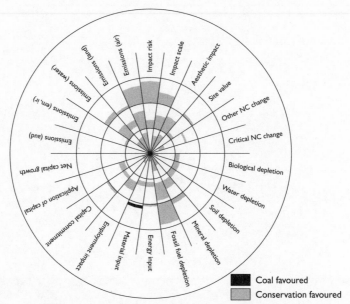

Figure 17: The comparison between a coal-fired power station and a major energy conservation programme

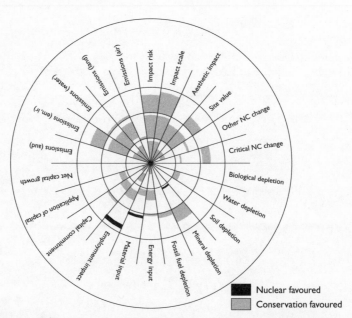

Figure 18: The comparison between a nuclear power station and a major energy conservation programme

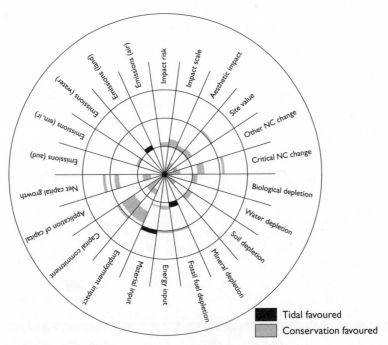

Figure 19: The comparison between a tidal barrage
and a major energy conservation programme

The SAMs presented here – with artifical data – would probably favour a decision to close the energy conservation programme. The important point, however, is that the SAMs that show the differentiated opinion profiles reveal most of the important trade-offs inherent in all of the possible decisions open to the planning authority. This greater transparency would in itself tend to promote a more accountable and inclusive decision-making process.

15

Conclusions and Recommendations

The global ecological crisis

The authors believe that the Earth is now being subjected to some of the most significant and extensive changes since human beings evolved. These changes are permeating almost all the systems on which human life depends, including the air that we breathe, the water we drink, the foods we eat and the physical resource base we mine. We are altering the conditions that shape evolution itself.

The processes that give rise to pollution, toxification, resource depletion, destruction of biodiversity and the anthropogenic chemical and physical changes to the atmosphere, the land and the marine ecology are, with few exceptions, increasing.

Many people believe that nature has to be protected from human depredation, that we live on a 'fragile Earth', and that the resilience of the systems concerned is being steadily reduced. The authors believe that this is inaccurate. Life itself has great resilience. Most of the thriving complexity of life today has developed since the mass extinction at the Permian–Triassic boundary, some 250 million years ago, at which perhaps 95 per cent of all species extant were destroyed. In general, therefore, these environmental systems adapt and endure. Their characters have, however, undergone a number of fundamental changes in the course of this history. This process of evolution has not stopped. It is possible, therefore, that the character of these environmental systems will change again, possibly in ways that would make human existence harsher and perhaps ultimately impossible.

While atmospheric changes, for example, may cause drought and flooding, and consequent disease and famine, the atmosphere itself remains and the web of life will adapt to the changes in the Earth's life support systems. It is the human species, with a long maturation period before fertility and a remarkably sophisticated but relatively fragile biological structure, that will be vulnerable to these changes. The human species has also taken a number of steps to resist some of the pressures of evolution, such as the creation of health care systems, which are arguably amongst our finest achievements, but which could decrease the ability of the species to adapt to any fundamental change in the global ecology.

One-fifth of the population of the planet – largely the populations of North America, Western Europe and Australasia, and the other powerful elites of the world – consume about three-quarters of the world's resources, influence or control most of the significant flows of both natural and artificial capital, and generate, either directly or indirectly, much of the pollution. This same global elite will be relatively protected from any changes to the global ecology. The wealthy can outbid the poor for increasingly scarce resources, have the sophisticated military capability to protect their interests, can afford to construct sea defences for their cities, and have the advanced technology and health care systems to afford themselves a degree of protection from increased levels of UV-B and toxic pollutants.

In terms of material well-being, personal and political freedoms, and education and overall level of health it is almost impossible to argue that the elite who live today in the wealthy nations are not better off than any previous generation or peoples. The authors believe that, despite this, it would be foolish even for the elite not to view the changes in the global ecology with alarm, even if only from a strictly self-interested viewpoint. No elite will be able to protect itself completely against a fundamental change in some global ecological parameter. At a less apocalyptic level, if some of the more worrying trends continue as at present, the world may become a harsher and more dangerous place, with fewer forests and more deserts, with whales, tigers, elephants, gorillas and many other forms of life existing only on film and perhaps as samples in a gene bank, with a human population of eight or ten billion or more, many scrabbling for some sort of life below subsistence levels, and with an ever-present risk of crime, disease, insurrection, and wars for resources and lebensraum. Even if the elite felt confident in their ability to survive these threats, they might question their ability to continue to requisition three-quarters of the world's resources on an indefinite basis.

It is also important to note that while increases in population in the poorer nations are a matter for legitimate concern, it is the compound effect of population level, per capita consumption level and the associated pattern of global resource demands of the rich rather than the poor that is currently creating most of the stress on the global ecology. Thus while the global ecological crisis is of global concern and is a global responsibility, the authors believe that that it is the rich nations who have the greatest share of the responsibility to respond to the challenge. It is also the rich nations that have the skills, resources and political and economic power to start to develop solutions to these problems.

Reasons for pessimism

For a number of reasons, the authors are pessimistic about the likelihood of an adequate response to the extremely serious changes that have occurred and continue to occur in our ecosystems. Some countries, governments, businesses and individuals have begun to take the global ecological crisis seriously, but progress is slow. There is a lack of urgency in the speed with which the problems identified are currently being addressed.

One important reason is, perhaps, that decision makers do not believe that the problems are as serious and wide-ranging as ecologists and others suggest. It is also possible that the rich and powerful of the world do not expect to suffer the immediate consequences of environmental damage. Perhaps a more worrying reason is that people who are not immediately and continuously confronted by the effects of environmental losses seem to resist the logic of any argument for change. There is little evidence that people are generally willing to vote for changes to their life-styles or make material sacrifices on the basis of marginally perceived threats.

Tragically, the spur typically required to lend urgency to these issues is an immediate threat or disaster. The startling discovery of the ozone 'hole', for example, led to remarkably rapid action to reduce the use and release of CFCs. Such individual disasters do not, in general, seem to lead to a systematic reappraisal of the nature of human interaction with the life systems that support us, still less to a continuing sense of urgency. There seems to be a rather fundamental tendency for people to become inured to threats. This seems to hold true even when the same people believe, objectively, that the threats may be increasing. People are willing to live beside volcanos, for example, and to accept the risks of living in earthquake zones, even when they genuinely understand the nature of the danger.

It is hard to avoid concluding that a necessary condition for sustained motivation for change may be repeated direct and tangible evidence of threatening environmental decline in the form of a frequent series of ecological collapses. It is quite conceivable, unfortunately, that such a condition will be met.

The final reasons for our pessimism are the lack of appropriate mechanisms for effecting change, and the power and degree of vested interests that seem likely to resist change. We may need new international institutions, agreements and understandings, for example, which implies at least some reconciliation of the various divergent interests and conflicts in the world today. It might also be necessary to curb the ability of multinational companies to evade environmental regulation by moving beyond the jurisdiction of nations in a position to impose such regulation.

Many of our proposals, and the proposals of others working in this field, will be strongly resisted with claims that the suggested change is unnecessarily radical and that the problems have been exaggerated, or that the change is not adequately justified by scientific studies. The authors believe that the reverse is the case, and that the proposals contained in this section would constitute a modest response, given the scale of the problems.

Commitment, analysis and action

This report does not present simple answers. This is because the concept of sustainability is not simple. One could make a clear and uncomplicated commitment to move towards a more sustainable way of life, and the actions that one would wish to take at any one point might well be quite straightforward, but the analysis that must underpin any genuine transition to sustainability, given the highly complex pattern of interactions between the social, economic and environmental systems concerned, must be sophisticated.

All three elements are essential. First, it is vital to build a wide consensus on the direction in which society should go. Given the scale of the challenge, this is likely to be a lengthy process. Perceptions, attitudes and expectations may all have to change.

Secondly, any programme for a transition to sustainability must be translated into a series of clear, practical and politically feasible steps. Given the fluid and dynamic nature of the challenge, it is unlikely that we will ever be able to anticipate exactly what steps might be necessary and practicable at all future points. What is important, therefore, is to construct a decision-making process that is adequate to the task. This decision-making process must have feedback mechanisms at each of the key levels of policy development and action. Broad policy commitments must be continuously translated into practical steps and the results compared with expectations, in order that the next series of steps can allow for overshoots and undershoots. The policy commitments themselves must evolve in the light of increasing knowledge of the social and natural worlds, and of the nature of the interaction between them, and in the light of the practical experience that will be gained during the course of implementing the strategy.

Thirdly, while our actions, ideally, will be simple, our thinking must be sophisticated. We must think carefully about the various dimensions of the problem, and about how they bear on each other.

Models as servants, not masters

The crux of the sustainability debate involves the nature of the relationship between qualitatively different types of systems operating in different domains. We do not have an agreed way, as yet, to capture the behaviour of this world of interacting complex systems, such as ecological and economic systems, that are generically different and adapt in different ways. It is possible that we will never have a single way of representing this order of complexity.

The authors believe that a systems approach is the key to the integration of the necessary analyses of the complex social, economic and environmental systems implicated in any significant development. A systems approach involves looking at the various environmental, economic and other systems with which we are concerned as systems, that is, examining them for attributes that are true of systems generally, focusing on the pattern of interactions and exchanges between them, developing models of their behaviour, and improving our understanding and our ability to predict outcomes.

It is important to note that a systems approach is a general perspective, and involves more than systems modelling. Systems modelling is an attempt to draw all the key dimensions present into a single model. This is, in general, a highly mathematicised approach, and is usually applied to relatively 'hard' and quantifiable systems. A more generalist systems approach, which can be extended to 'soft' systems that cannot be quantified on an equivalent basis, emphasises the development of an understanding of the pattern of interaction between the systems concerned. This usually involves drawing on a range of models and analytical tools and constructs developed in the various specialist disciplines involved. The role of a general systems approach is to provide a perspective and a context for assimilating and integrating the insights offered by different techniques.

With a systems approach, models are an important aid to understanding, but they (and other such techniques) are valuable only insofar as they can allow us to take cognisance of a far greater range of variables and interactive effects, and so improve the quality of our thinking. It is important not to place too much reliance on system models, or economic models, or on any other kinds of models. When dealing with complex adaptive systems, and where there are unknown areas of uncertainty and indeterminate behaviour, we can never be completely certain that any model is truly adequate.

This is not just a technical point. It is of fundamental importance. The implication is that we must accept limits to our understanding and control. This in turn implies that we must move beyond closed systems thinking and abandon the search for absolute knowledge and for a point of certainty on which to construct an edifice of scientific knowledge or political ideology. The real challenge of sustainability is to move to open systems thinking – pragmatic, ethical rather than ideological, and beyond simple certainties.

The precautionary principle

It is rarely possible to say with any degree of certainty what the effects of various actions on the global ecology will be. The systems involved are too complex, too interconnected and too highly non-linear for high degrees of confidence. This is not to underestimate the truly impressive understanding that we do have of much of the natural world around us, particularly in the physical as opposed to the biological or social realms. It reflects, rather, the interlocking nature of ecological systems and the unprecedented scale of changes. In many cases such changes stretch our theories into previously uncharted territories. No matter how accurately we can predict the outcome of a specific action, one interaction with one system can, under some circumstances, trigger a cascade of effects in a number of other systems. Natural evolution has produced all the beauty, complexity and delicacy of that which we see around us through contingent processes. There have been some dramatic perturbations and major discontinuities, such as mass extinctions, during the course of evolution to date. Nevertheless, the balance of life that we see is a dynamic near-equilibrium, an astonishing mosaic of largely self-adjusting processes and systems that together give rise to the web of life of which we are a part and on which we depend. These systems are characterised by the remarkable intrinsic robustness inherent in any assembly that has survived the completely ruthless selection of evolution and adaptation in a changing world. There are, however, demonstrable limits for any given system beyond which it cannot survive. Given our limited understanding, we can rarely be confident in predicting these limits. In many cases we have probably not even recognised the existence of a system or regulatory mechanism that was stressed to the point of rapid change or collapse.

Taken to extremes, this understanding of the world as a complex interlocking web of systems that are, individually and collectively, vulnerable to gross interference might be used to justify complete stasis, and a reluctance ever to contemplate or test any change of practice. The authors would like to stress that this is not their position. We would suggest, however, that when changes being made are on a particularly large scale or radical in nature, there should be a rather greater emphasis on forethought, research and planning. Critical systems should be presumed vulnerable, rather than the more typical presumption that they are resilient or invulnerable. This is the essence of the precautionary principle.

Adoption of the precautionary principle results in an increased emphasis on the need to demonstrate the safety of any proposed new product or process, for example, before it is released into the wider environment. The precautionary principle can be developed, in various different contexts, into specific regulatory procedures. It has a more important role, however, as a way of encouraging a wider consideration of the side-effects and wider implications of apparently sensible actions. Rather than assuming that chemicals used to kill pests on crops are otherwise environmentally harmless or neutral, for example, the precautionary principle might suggest that we should consider the possibility that a chemical harmful to one pest might cause a range of other ecological consequences. Similarly, where the theory of substitution in neoclassical economics indicates that the scarcity of one resource will lead a rise in its price and so to the timely development of substitutes, the precautionary principle might lead one to consider the possibility that resources are not always commensurate and substitutable. This might then encourage a search for lateral solutions, which might be, for example, to encourage an existing predator in the first example, and to take measures to control the consumption of the resource in the second.

It is important that the precautionary principle is not seen or used as a way of discouraging all development. A great deal of further scientific research and technological innovation, for example, will be essential if we are to develop more energy and resource-efficient modes of production, feed the world's growing population and so on. The precautionary principle would, preferably, represent a component of a move towards a more comprehensive and intelligent appraisal of the wider role and implications of particular changes and trends.

The goal of sustainability

It seems unlikely that it will ever be possible to construct a fixed set of social and economic arrangements that would be permanently sustainable in environmental and other terms, given that the world itself changes and evolves. Species develop, flourish and perish, the composition of the atmosphere changes, and the dynamically-interrelated network of relationships that constitutes the global ecology gradually transforms over time, thereby changing the ecological parameters within which the human species has to operate. The evolution of biological and ecological systems on this planet was and is a contingent and heuristic process. It is unlikely that the appearance of the human species, for example, was an inevitable outcome of the process of evolution, just as the continued survival of the human species is in no sense guaranteed. There have been a number of points in the history of this planet at which events could probably have taken a different turn, and there will probably be many more such points in future.

It is impossible to eliminate all risk in such a contingent process. It is probably more accurate, therefore, to think in terms of reducing rather than eliminating the overall risk to which the human species might be exposed, and of reducing the number and impact of activities agreed to be unsustainable rather than aiming for a definable state called sustainability. In order to do this, it will probably be necessary to control particular activities, by restricting actions, for example, that place undue pressure on particularly sensitive or critical ecological functions.

At any one time, therefore, there will be a particular set of social and economic arrangements that, on the basis of the best available information and analyses, offer the best prospect of continued viability. These social and economic systems are soft systems, because many of the basic structures are matters of belief and definition. They too change and evolve over time.

The key to achieving sustainable development, therefore, is to understand and shape the interaction between complex adaptive natural systems and soft socio-economic systems in order to ensure that we always remain within our survival region at the intersection of the survival regions of all the systems on which we are dependent. Of course, the nature of this interaction between natural and socio-economic systems is itself constantly evolving, as species regenerate or become extinct, resources are exhausted or new reserves discovered, social and economic systems expand and collapse and new technologies are developed, disseminated and superseded. Thus both problems and solutions are dynamic.

The task is so complex that it is unrealistic to aim to respond to every minor fluctuation in the natural environment, or to control the details of operations at every level of human society. A more practicable strategy would be to ensure that the systems concerned remained within the key parameters while allowing their detailed internal dynamics to self-adjust.

A government might set an overall carbon emission target, for example, then leave it to industry to find the best way to achieve that target, or a government agency might set a quota for the tonnage of fish to be taken, then allow the trawler-owners to bid for their share of that quota. Of course, this approach does require some mechanism to enforce targets and quotas, with appropriate penalties, such as fines, for failure. To be effective, such penalties would have to be more expensive than the cost of compliance.

Similarly, an efficient way of ensuring that wastes were eliminated or recycled would be to raise the price of virgin raw materials and energy. It might be necessary to prime the market by making sure that people were aware of the opportunities to recycle, and of the existence of sources of recycled material, but market processes would ensure that people then changed their behaviour and recycled materials that would have become too valuable to waste.

The specific recommendations in the rest of this section are some suggestions for action proposed by the authors along the lines indicated above. They are all changes that we believe would be generally helpful in raising awareness of the underlying issues, addressing some of the more dysfunctional outcomes of the existing social and economic arrangements, and in moving society towards a more sustainable way of life. Some of these suggestions are concerned with political economy, such as measures to make the distribution of resources between generations and between peoples more equitable and just. Others are concerned with reducing the impact of economic operations, such as improving energy and resource-use efficiency, developing cleaner technology, minimising waste, enhancing yields and reducing waste and pollution. The authors believe that this combination of measures would simultaneously reduce the burden on the global environment and extend the life of remaining resources, thereby giving us more time to explore options for a permanent transition to a more sustainable way of life, while also starting to address some of the fundamental reasons why the problems arose in the first place.

Global initiatives

Most unsustainable practices have an intrinsically international character. This is partly because environmental effects do not respect borders, partly because the existing international imbalance of power is deeply implicated in many unsustainable practices, and partly because it is unrealistic to expect countries to act against their immediate perceived interest for a global good without coordinated action from other nations.

This suggests that there is an urgent need to establish a set of international bodies, perhaps under UN auspices, to be responsible for creating and policing standards governing those practices that are of international character and global significance. Such a body would need to carry the force of international law to be effective, and would have to be fully democratic and accountable to ensure support.

The authors are under no illusions about the difficulty of establishing such organisations and making them effective, but at the moment we can see no feasible alternative. The call for a global tier of authority to complement national and local governments represents little more than a recognition of the reality that many of the important industries and businesses are multinational, and that many of the resource-use and pollution issues are truly global, so that no one government or even group of governments can hope to exercise adequate control over either.

It is equally important that the extreme global inequalities of wealth and power are also addressed as a matter of international concern. Programmes to eliminate the debts of the poor nations and to level the terms of international trade, and the imaginative and creative co-development of more sustainable development paths for the poor nations would, in combination, solve many of the more pressing social and environmental problems in the developing world.

There are a number of existing mechanisms and organisations that could be adapted, developed and extended to play key roles in this regard. It would be possible, with sufficient political will, to change the remits of the International Bank for Reconstruction and Development (the World Bank), its subsidiary operation the International Development Association and the International Monetary Fund, and to re-orient them to more socially and environmentally sustainable development. The World Trade Organisation could play a constructive role in ensuring fair and environmentally-sensitive terms of international trade, but this too is only likely to happen with sufficient political will, vision and guidance.

The existing mechanism for addressing regional disparities in levels of development within the European Union is a model that could be extended globally. All European Union members contribute funds to and withdraw funds from a common exchequer. The developed member states contribute more than they withdraw, while the less developed states withdraw more than they contribute. The funds are transferred to the less developed states in the form of regional development funds, for example, which constitute grants for infrastructure development projects. Over time, such directed funding is intended to bring the level of development in the recipient states closer to the average. A global development fund, in conjunction with socially and environmentally-sensitive regional development funding, could play a similar role for the world community.

The final global imperative would be to ensure the universal availability of birth control. The authors believe that this should be done on an voluntary and non-coercive basis. The evidence suggests that a degree of economic security combined with measures to raise the status of women, along with the provision of educational facilities (particularly for women and children) and the introduction of measures to restrict some forms of child labour generally provides an environment in which women and men choose to stabilise families around replacement rates.

The role of government

The authors believe that the UK government has to take the lead in any UK transition to a more sustainable way of life. Local authorities, for example, can do little without the necessary permission and support from government. Similarly, any one firm, no matter how responsible and committed, can only go so far without the necessary lead from government. Responsible businesses, that internalise their environmental costs and make due provision for environmental restitution, may be undercut by less responsible or scrupulous competitors. Governments, on the other hand, can raise the required standards for an entire sector, thereby ensuring that the responsible companies prosper and the less responsible reform.

Economic indices

The authors believe that the department of the UK Government with the greatest potential ability to effect a transition to a more sustainable way of life is the Treasury.

One significant reform, for example, would be to abandon the use of traditional forms of accounting as the sole basis for the formulation of economic policy. The central conventional measure of economic well-being, gross national product, is unhelpful in that it has no regard to the composition of the product. GNP increases when shoddy goods are produced and then replaced, and when buildings collapse and must be rebuilt, but decreases when durable goods are produced and then properly maintained. Similarly, GNP increases when finite resources are extracted, but reduces when resources are preserved or used efficiently. Human and environmental benefits, however, seem to accrue in the opposite direction.

It would be preferable, therefore, if the Treasury were to take a much wider array of economic indicators into account, including indicators of environmental change and resource flows. This will in turn require the development of decision-making processes that utilise multiple indicators rather than a single aggregated figure.

Managing the economy

The authors believe that a transition to a more sustainable way of life will be best effected through an appropriately regulated market system. Command economies have failed for a number of reasons, including their relative inflexibility, poor decision-making processes, and failure to motivate participants. Market economies have far greater flexibility, partly because the decision-making process is largely decentralised to the individual market operators.

Markets, however, have their failures too. For example, markets only operate efficiently if market operators have the information they need to make their decisions. In the real world, however, we must deal with very uneven information; sometimes the information is too complex or the amount of information is too great to be readily understood, and sometimes the necessary information is unavailable. Humans are also imperfect processors of information, and rely on assumptions, heuristics and precedents as a basis for making many decisions.

There are also factors which will tend to move a market system away from sustainability. The profit motive, for example, tends to provide an incentive to externalise as many of the costs of production as possible. This could mean operating with minimal environmental controls, or paying the lowest possible prices for raw materials, and to avoid paying any wider environmental costs that the use of these materials might entail.

If a market system is to become sustainable, therefore, the appropriate information must be introduced into the market. If the use of any particular material has some potential long-term or wider social or ecological cost that was not captured in the market price (as opposed to simple scarcity, which one would expect to see reflected in the market price), this must somehow be reflected in prices in order to send the appropriate signals to consumers. In this way economic actions can be connected to their environmental consequences. Of course, many of the costs with which we are concerned will happen in the future, or in some other part of the world. We must therefore find ways of bridging these gaps. This will involve making some large estimates as to the extent of future or displaced consequences, and then calculating the appropriate level of price change to send the necessary signal.

These operations would be a form of market correction. The authors do not believe that an unregulated market system, without such corrective mechanisms, would operate within the parameters of sustainability. Market prices will reflect current resource scarcity, but cannot reflect the ecological value or the future value of a resource without some process for estimating and incorporating these values in prices. The market itself could not arrive at the final prices, so any externalised or future costs would have to be calculated and then imposed in some way on the market. The market would then move to a new balance between supply and demand that reflected the relative changes in the costs of materials.

The authors believe, therefore, that there will have to be appropriate macroeconomic planning and market regulation if society is to combine a reasonably high level of efficiency in resource allocation with long-term observance of critical environmental and social limits and goals.

There are a number of fiscal and regulatory tools available, many of which have something to offer in this regard. We have listed a number of suggestions in the remainder of this section, which are not mutually exclusive. One approach might be useful in one context, while another approach might be more appropriate elsewhere.

❑ One approach would be to establish processes for valuing and charging for natural capital, then utilising an otherwise conventional cost–benefit analysis with a unidimensional numeraire (probably, but not necessarily, expressed in terms of monetary values). The object of this exercise could be, for example, to maintain the total economic value of the combined sum of artificial and natural capital, allowing at least some substitutions of artificial for natural capital. The necessary calculations of economic value-equivalence could be done for each individual commodity, by calculating the appropriate charges for different kinds of timber, fish and so on, then imposing the appropriate tariffs, or could be done and then implemented on a generic basis by introducing basic energy and resource-use taxes and pollution charges.

The advantage of this approach is that it represents a relatively straightforward correction of an obvious market failure, and can therefore be accommodated within the current economic paradigm. It builds in an essential missing feedback loop by improving information in the market as to the real costs of natural capital, and would reduce the excessive and sub-optimal use of resources that invariably follow from the undervaluing of natural capital.

This approach would, however, be limited by the extent to which it is possible to derive realistic values for natural capital, which in turn is limited by the non-equivalence of human and ecological values, and the extent to which it is possible to make substitutions of artificial for natural capital.

❏ A second approach would be to establish compensating or 'shadow' projects that would attempt to compensate in ecological or other terms for the loss of natural capital. For example, a government agency could provide an environmentally enhancing (but probably 'uneconomic') project that compensated for relatively unconstrained environmentally damaging market operations.

The advantage of this approach is that it would allow markets to attempt to produce the economically optimal allocation of resources without the need for additional complex calculations of the costs of the environmental damage associated with each individual development (although this could also be achieved by the introduction of basic energy and resource-use taxes and pollution charges), as environmental costs could be compensated in appropriate environmental terms. This approach could also be accommodated within the current economic paradigm, and may be more politically feasible than the first approach above.

The main drawback of this approach is that there are limits to the extent to which 'enhancing' projects can genuinely compensate for damaging developments, an idea that makes more sense in economic terms than in ecological terms. We cannot manufacture new land, and we cannot enhance virgin rain forest, for example, so the only land that could be genuinely enhanced is land that was previously damaged. Unless damaged land can be restored to equivalent or greater value than the virgin land that was sacrificed – which in many cases is unlikely – more damaged land must be restored than virgin land consumed if we are to maintain some sort of total ecological value. This might well be desirable, but leads back into political difficulties as the cost of a fully adequate restoration may well be greater than the profit margin of many proposed developments. From a purely technical point of view, this is not a strategy than could (or would have to be) pursued indefinitely, as eventually one must run out of damaged land to restore, even allowing for the fact that one could then start restoring the land damaged in the interim. More seriously, this approach also leaves open the question of how to deal with the inherited burden of historic environmental damage. To meet these historic costs implies restoring all damaged land, which therefore leaves none to be restored as compensation for contemporary projects. The most fundamental limit, however, is that irreversible losses (such as species extinctions), and energy conversion costs mean that it is unlikely in practice that all land could be truly restored to its full pre-exploitation value.

❑ A third approach would be to establish through either scientific enquiry or political debate some limit to be set on the use of natural capital, with tradable pollution permits and tradable resource-use permits then used to allocate those resources optimally within the set constraints.

The advantage of this approach is that it avoids all the valuation problems involved in trying to convert some ecological function into a human value system. Ecologists rather than economists would determine ecological values, atmospheric scientists rather than economists would determine appropriate carbon loadings and so on. Society would have to decide whether to observe higher or lower limits than those recommended. Once such limits had been decided, however, the management and allocation of the total ecological resource thereby defined would become a relatively straightforward economic question.

The drawback to this approach is that it could be overly restrictive. It would be possible to have greater flexibility in the way and extent to which natural capital could be used if compensatory projects could be used in some instances to make good the costs incurred by particular developments.

❑ Thus a fourth approach would be to use elements of all three approaches listed above, allow compensatory projects where appropriate, and to derive an appropriate sequence of policy measures, such as in this example;

1   The first step would be to decide on the degree of criticality of different forms of natural capital, then to place those forms of natural capital believed to be most critical off-limits by whichever regulatory or fiscal policy instrument was deemed most appropriate at the time.

2   The second step would be to determine annual limits for the use of non-critical natural capital, then to use appropriate resource allocation measures (such as tradable resource use and pollution permits) to allocate those resources.

3   The third step would be to institute public sector provision of appropriate and commensurate compensating projects as an allowance for the loss of the non-critical natural capital. Examples

of appropriate compensating projects might include the restoration of contaminated land or the reafforestation of denuded land. More sophisticated forms of compensation would include the planting of trees to absorb carbon and in that way compensate for the burning of fossil fuel, or the use of fossil fuel revenues to fund the development and deployment of renewable energy technologies that could eventually displace fossil fuels as an energy source.

This approach would be powerful and flexible, and could be accommodated within a social market framework.

It is possible, however, that this approach would not give sufficient weight to historic costs and damage, or to the cultural codes and norms that may constrain the range of possible developments and compensating projects.

❑ A systems perspective, therefore, suggests a still more comprehensive approach than the fourth approach above. The fourth approach would form the basic strategy, but, drawing on the conceptual tools of institutional economics, social, psychological and cultural factors could be brought into the same decision-making process. This would require using non-equivalent indices and modelling techniques, as described in Chapter 14. The authors believe that this would give the best prospect of long-term sustainable development because more of the actual decision-making process would be revealed, and the current uncertainties surrounding these issues make it important to make the decision-making as open-ended and as transparent as possible.

Legislative and fiscal change

The following reforms are recommended on the basis that they would encourage more responsible corporate behaviour, encourage industry to become increasingly energy and resource efficient and less polluting, and enable the provision of the finance necessary for a long-term reconstruction of the industrial base of society. It should be noted that it would be necessary to change the current culture of the fiscal and banking system as well as restructuring some of its operations in order to effect the necessary changes.

One of the most useful general reforms would be to insert some of the missing information feedback loops between investment decisions and corporate actions and the environmental and social consequences. This would entail clarifying and where appropriate extending the responsibilities of directors, shareholders, investors and bankers, and emphasising greater transparency and accountability in corporate behaviour.

This could be done in various ways, such as in the following examples:

❑ If company directors were made more legally accountable for environmental or other disasters caused by corporate neglect or culpable behaviour this would undoubtedly encourage a greater emphasis on environmental and related safety issues at board level.

❑ If the costs of remedying environmental damage were firmly attributed to the corporation that caused the damage, then investors and bankers would undoubtedly pay more attention to a firm's environmental behaviour and would be more likely to insist on proper reporting of environmental performance.

❑ If company auditors were obliged to report evidence of corporate evasion of environmental regulation, as they are evidence of financial impropriety, this would make environmental regulation more effective by ensuring that more cases of malfeasance would come to light.

There are a number of further measures that would make it easier to raise the money for long-term investment, encourage innovation and improve efficiency in the productive sector of the economy, especially if done in conjunction with the steps outlined above. For example:

❑ Tougher penalties for breaches of environmental limits would shift the balance of costs and benefits in favour of greater investment in clean-up operations, waste-minimisation techniques and cleaner technology.

❑ The introduction of taxes to make energy and resources more expensive would make it more advantageous to invest in energy and resource-use efficiency measures.

❑ Tougher penalties and more expensive inputs would put pressure on the less enlightened firms to comply with environmental standards, and would help to prevent them from competing unfairly with those firms that behaved responsibly and internalised more of their environmental costs. They would also make it harder for the less enlightened firms to maintain an unrealistically high rate of dividends, as more capital would have to be diverted into clean-up operations. This would make it easier for the more responsible firms with high environmental standards and high rates of spending on research, development and reinvestment to compete for and retain investment.

❑ More long-term investment could be encouraged by establishing specialist industrial credit banks, constructing institutional links between such investment banks and industry, and making long-term debt more attractive with differential interest rates and more feasible by encouraging long-term deposits.

❑ Industries could also be encouraged to finance reinvestment and renewal of their capital base from profits rather than borrowing, and through borrowing from investment banks rather than through share issues, which would help to reduce the pressure for short-term returns. This could be assisted by giving appropriate tax breaks to those firms that accumulated financial reserves, provided that these were genuinely used to fund research, development and innovation.

Finally, on a macroeconomic level, various measures could help to shift the balance from consumption to investment. This could be done by, for example, introducing credit controls and controlling credit growth, increasing investment credits and restricting consumption credits. Consumption credit controls could be relaxed in order to stimulate demand and so absorb production if necessary when the economy was at the low point of a cycle.

The role of industry

In the light of current knowledge of environmental and related impacts and trends, it is a reasonable assumption that a society that prioritises the production of essential goods, produces these with ever-cleaner technology, and operates to increasingly high standards of energy and resource-use efficiency will make a greater contribution to global sustainability than one that fails on any of these grounds.

As indicated earlier, this is unlikely to provide the whole solution to the problem of sustainability, as there are important political, economic and ethical questions as to how resources should be used, shared and distributed between generations and between nations. The move to cleaner and more efficient modes of production will not be a sufficient condition, but it will be a vitally necessary condition.

It will be necessary, for example, to develop and implement more fundamental solutions to pollution problems, and to move towards eliminating pollutants where practicable rather than seeking to disperse and dilute them in some appropriate environmental medium. It is now recognised that, in some cases, strategies of dispersal and dilution have resulted in environmental impacts being displaced or accumulated elsewhere. The construction of tall power station and factory chimneys in the UK, for example, resulted in a local improvement in air quality but also in increased rates of acid deposition in Scandinavia.

Similarly, the application of environmental controls to European rivers has failed to solve the problem of the cumulative accretion and biological magnification of toxins in the North Sea, which receives the discharges of a number of major European rivers and therefore forms the ultimate sink for a number of pollutants. National agencies are, understandably, concerned primarily about the rivers that flow through their jurisdiction rather than the control of aggregate pollution in the North Sea.

More fundamental solutions will probably involve cleaner technology and possibly industrial symbiosis in order to push towards waste minimisation and even zero emissions at either plant or site level. The future emphasis of government action should be in the direction of greater local responsibility, so that pollution problems, for example, are solved at or near the source rather than being displaced across geographical boundaries or down disposal routes until they accumulate and reappear in the final sink.

The role of business directors and managers

Developing a strategy

It is not possible to lay down a fixed plan and timetable for a transition to a sustainable way of life. What is needed is an open-ended, flexible approach. This makes it particularly important to appraise decision-making processes, and to ensure that they are able to cope with uncertainty. A transition to sustainable policies and practices requires long-term commitment, investment, reorganisation, and training. This means that a rather more comprehensive approach is required than with more usual organisational planning and development issues. This kind of change will entail periodic reviews of the organisation's definition, mission statement and relations with other organisations.

Of course, any one organisation has limits to its jurisdiction, influence, or control. Some of the most important parameters in a transition to sustainability will require changes in the planning system and fiscal system that could only be made by a central government, or by the pan-national institutions of an organisation such as the European Union. Similarly, no one organisation will be in a position to effect all of the suggestions listed in this section; these are given in order to provide a fuller illustration.

Policy assessment. The first step is to introduce a set of general parameters for assessing policies, projects and activities. No such set of rules could be a complete set of guidelines for sustainability. They are indicators of basic changes which, in the light of current knowledge, would probably help to move society in the direction of a greater degree of sustainability. Any change, for example, in the direction of greater efficiency in the use of energy and resources or towards reducing total consumption is likely to be helpful in this regard.

Managing change. The second step is to conduct an internal appraisal, and to look at the organisation's role, structure, policies, strategies and decision-making processes. It is important to introduce effective mechanisms for monitoring change in key areas and for assimilating this information into an evolving strategy. A great deal of relatively unimportant information must be excluded from any decision-making process. However, at times of changing demands, it is important to check whether the process of decision-making is excluding information that has become relevant. Compartmentalisation (both within and between organisations) is particularly problematic, as narrow remits are not appropriate when dealing with large and complex issues. Eventually, it may be appropriate and consistent to consider internal change, such as utilising non-material systems of reward and status recognition and personal development techniques to reduce the average per capita consumption of personnel. Internal education and training, with an emphasis on environmental education and systems thinking, will be particularly important.

Financing projects

In addition to the above, it is likely that, in the prevailing economic order, it will be necessary to introduce special conditions for projects that are being assessed for sustainability.

The additional conditions are as follows:

❑ Some developments, particularly those with a high social or environmental value and especially where some form of critical natural capital is concerned should be assessed on a zero discount rate basis. With some applications, particularly where compensatory developments are involved, it may be appropriate to discount both costs and benefits.
❑ A sustainable development option may require subsidy, possibly by being permitted an extended phase of investment, then should be expected to move into profit only after correctly internalising costs and setting aside adequate funding for reinvestment.
❑ It may also be appropriate, within the context of a larger strategy, to develop projects that did not make a financial profit, but which compensated for environmental costs incurred elsewhere.

The role of the planner

Planning agencies

Any organisation that has a statutory, advisory, consultative, or other input into planning processes can be effective in a number of ways.

1   *Advocacy.* This could include promoting the idea of sustainability to other organisations, and helping to educate the public. A transition to a more sustainable way of life is likely to require a degree of change in understanding and attitudes. This process can be fostered by supporting developments in environmental training and education.
2   *Planning guidance.* This would require promoting appropriate changes in structure and local plans, and the introduction of relevant principles in the assessment of planning applications.
3   *Changes in internal management processes and operations.* Most organisations occupy premises, and are purchasers, employers, and polluters, which means that there is scope for change in internal management practice. Such changes can be effective in a number of ways. For example, environmental conditions applied to purchased materials feed back through the economic system against the flow of resources, and eventually affect primary sector producers and secondary sector processors and manufacturers. These decisions, therefore, eventually affect a much wider community. Such changes in practice should be

based on regular and systematic environmental audits.

4   *Providing and encouraging remedial works.* There are instances where public sector organisations can contribute to or cover the cost of environmental damage that was at least partly caused by private operations. However, this could be complementary to the 'polluter pays' principle, rather than an alternative to it, which requires a little more explanation. As outlined in Chapter 7, goods and services in a market economy will be produced up to the margin, the point at which the value of any further production is less than the total cost to the producer of obtaining that additional production. It is also true that this total cost to the producer will be less than the social cost, as some of the costs of production will not be borne by the producer. This inevitably leads to a misallocation of resources and therefore a sub-optimal outcome for society. The polluter pays principle has the effect of internalising some of the costs of production, thereby correcting part of the misallocation of resources, which actually improves economic efficiency. However, while there are cases where it may be relatively easy to enforce this principle, there are others where it is very difficult or impossible. For example, it will clearly be technically difficult to attribute correctly all sources of pollution. This is likely to leave a need for public works, possibly in the form of compensating projects, to redress such generalised damage. Furthermore, previous generations have left an inheritance of degradation and pollution, and this too may have to be dealt with. This could be done either by assigning the problem to the current owners (of a contaminated site, for example), or by using public funds for such remedial work, or by some combination of these two policies.

5   *Promoting access to environmental information.* This would be a particularly important contribution, for various reasons. One is that the development of a pro-active approach on environmental issues will require some change in corporate cultures. A more open and accessible approach to business and government would assist this process. In the US, for example, much environmental information about companies and products is available on-line, in such services as the Toxic Reporting Inventory. This makes it easy to check on the environmental performance profiles of local industries. This in turn provides an incentive to those companies to improve their performance, and to build stronger relationships with their local communities. Local communities benefit by gaining a greater understanding of the role and impact of the forces that shape their environment.

All of these effects could be strengthened with planning and cooperation. For example, if several public sector organisations with related operations were prepared to review their purchasing criteria, they might find it useful to coordinate this process, to share information, and to consider coordinating purchase schemes to promote the development of the necessary markets. Similarly, a pooling of information and experience could develop into an approach to central government with a series of constructive suggestions for enabling legislation to permit further developments in this area.

Of course, an integrated local, national and international framework for action would make these contributions much more effective. Ideally, there would be a vertically and horizontally integrated hierarchy of planning levels, with each policy framework nested in a larger framework.

Horizontal integration (coordination between agencies or departments at approximately the same level) is the way to ensure that the overall strategy is carried out in a coherent and integrated manner, with consequent benefits in terms of efficiency and economies of scale, fewer errors and wastage, and more opportunities for joint actions and synergistic gains. Horizontal integration also conveys a more convincing and effective message to the wider community.

Vertical integration (coordination between agencies and departments at different levels in a single hierarchy) would ensure that local goals and actions were connected to regional, national and global results and vice versa, with consequently enhanced effectiveness and purpose. Global accords, for example, could form the basis from which national targets were derived. These in turn would inform regional and local plans, which would help to inform the decisions made by individuals. Such a framework does not yet exist. Action is still possible at all levels, but it will be less coherent and effective in the absence of an integrative structure.

The best hope for immediate progress probably lies in improving the coordination between local and central government planning. In Denmark, for example, there are effective national planning frameworks that allow national and local government administrations, agencies, research institutes and the business community to coordinate and cooperate on initiatives. This has allowed Denmark to make relatively rapid progress towards, for example, greater energy efficiency. Denmark decided to reduce its high level of dependence on imported oil after the shock of the oil price rises in the early 1970s by raising energy efficiency and productivity, and by developing renewable energy sources and combined heat and power schemes. The approach adopted included changes in the regulatory environment for the electricity generating industry, public investment in a heat distribution infrastructure, tax credits for relevant investments, the creation of a favourable set of local planning guidelines and conditions, directed funding for appropriate research and development, and support for comprehensive information, advice and other support services. This kind of comprehensive and integrated approach helped to convey a consistent message to all members of the community, and created a stable environment for research, development and investment.

Policy and project assessment

Much of the theory and technology needed for a transition to a more sustainable way of life probably already exists, at least in embryo. The most urgent task is to encourage changes in awareness and to enable the development and implementation of appropriate changes in policies and in practices. The precautionary principle, which is gaining increasing acceptance amongst planners of various kinds, forms an excellent starting point and could be usefully extended to many different kinds of development.

In general, planning for sustainability will require a better match of resources to needs, a much clearer definition of needs, and a more realistic appraisal of the implications of any proposed land use or related development in terms of the associated demand for resources. It is essential to have a comprehensive policy programme, and to develop a perspective across the breadth of the organisation's activities. Otherwise projects can conflict. One local authority, for example, found that its successful paper recycling scheme reduced demand for virgin pulp from the local paper mill to the point where it undermined the economic viability of that same authority's native woodland regeneration project.

With that in mind, there are a number of factors that should be considered by organisations when reviewing any development proposal.

Demand management

The first option should be to consider whether the demand for an activity or process could be managed, rather than assuming that demands should be met by making more resources available. For example, demand for transport and consequent road congestion might be addressed by development of integrated and more self-contained residential and work areas to reduce the total demand for transport, rather than the construction of more roads. This could be followed by measures to shift the balance of advantage from private to public modes of transport, such as traffic calming, road pricing, and the introduction of exclusive bus lanes and the construction of light rail systems.

A request to build a new water reservoir might be approved only after leakages have been eliminated, and demand control measures (such as water metering) considered. A request to open a new quarry might be approved only after the scope for recycling existing stocks of the mineral and the options for controlling demand have all been properly explored and exhausted. In such cases it may be appropriate to have regard to the end-use of the mineral. The mining of minerals for road-beds, for example, may be inconsistent with overall policy in this area, but could be allowed if the minerals were necessary for house-building programmes. Demand for sites for new housing and workspace should be encouraged towards derelict, brown-field, or gap sites to the maximum extent possible. This would increase urban density, which would make public transport and recycling schemes more viable as well as minimising the loss of green-field sites.

Meeting needs

The preferred development option would be one that also contributed to meeting the wider needs of the community. This will call for some careful thinking as to what these needs are. The expressed preferences of the individual members of a community will not necessarily give rise to the preferred option for the community as a whole, which is why we have governments and planners.

Take the example of transport. Various studies suggest that most people make many economic decisions on emotional at least as much as on rational grounds. Cars, as with such items as clothing and cigarettes, are sold partly on the basis of styles and images with which people identify. Some cars are sold with images that convey impressions of wealth, power and sexual success.

Furthermore, values change over time. In a society that is growing richer, goods (such as cars) that are considered a luxury by one generation can come to be regarded as basic necessities by the next.

There is no reason why such processes will automatically give rise to the optimal transport strategy for a city. The task for the city planner, however, is to ensure that people and freight can move quickly and efficiently through and around the city. Having defined the task in this way, it is then possible to decide whether this is more likely to be achieved with a public transport system than with private car use.

It is important to note that the meeting of individual and community needs will not necessarily be achieved by the promotion of economic development per se. It calls for the making of judgements as to the quality and net worth of the development, and the extent to which the development contributes to the overall strategy.

In a transition to a more sustainable way of life, all proposed development options should be analysed to clarify their contribution to real individual and community welfare. The preferred development options would be those that achieved any or all of the following:

❑ The direction of capital into directly welfare-related activities, such as the high-quality production of essential goods or the development of human potential.
❑ The reduction of welfare inequalities within the area concerned, possibly by utilising non-material systems of reward and status recognition to reduce the average per capita consumption of resources among the wealthier members of the community. The reduction of disparities in income and wealth, and consequent reduction of disparities in welfare could also be assisted by, for example, siting labour-intensive developments in areas of low income and high unemployment.

Contributing to sustainability

Certain developments have the potential to contribute to sustainability in a wider area. Some may entail some environmental or other costs, but may make a net positive contribution to sustainability. These are some of the general points to consider:

❑ Will the development result in a net reduction in the demand for resources? The construction of a new rail system, for example, would require resources, but could still result in a net reduction in energy demand if enough people transferred out of their cars. It is important to recognise that such gains can be lost if people respond to improved efficiency and consequent reductions in cost by increasing their consumption. The fuel savings made by using more efficient car engines, for example, would rapidly disappear if people responded to the reduced cost of their motoring by using their cars more. Similarly, the fuel savings made by insulating houses could disappear if people used the money that they had saved on their heating bills to pay for a flight abroad. It is important to ensure, therefore, that any moves to reduce demand are widely understood and supported.

❑ Will the development result in an increase in long-term sustainable net yield levels? Planting and managing woodlands, for example, will increase the sustainable yield of timber. Similarly, it would be possible to encourage types of agriculture that do not require high levels of fossil-fuel derived inputs, such as pesticides and fertilisers, and which are therefore more likely to be sustainable than types that do require high levels of such inputs.

❑ Will the development allow a reduction in imports of natural capital, such as minerals and timber, especially in those cases where those imports have high environmental costs in the countries of origin? This may be achieved by developing indigenous substitutes.

❑ Will the development allow a reduction in exports of natural capital, especially those that have high environmental costs? This may be achieved by developing economic alternatives.

❑ Will the development lead to greater efficiency in the processes that convert natural capital into artificial capital, such as in the construction of buildings and machinery? It is important to remember that such gains in efficiency should be calculated over the entire production process, so that one apparent gain is not offset by a greater loss incurred at another stage in the process.

Other points to be considered are reviewed under the two headings of materials and energy.

Managing the demand for materials

The preferred development option would be one that achieved one or more of the following:

❑ The development of methods for recovering and recycling materials within the economic system and reducing wastes and dissipative losses. Construction materials, for example, could be collected and re-used to a much greater extent. Another idea with great potential is to develop 'closed cycles', where the waste output from one process is used as the input for another. Sewage, for example, can be converted via methane digesters into fuel, water and soil conditioner. Even more ambitious is the idea of 'industrial symbiosis', where industries are appropriately zoned and linked together so that the waste from one industry forms one of the inputs for the next industry and so on.[115]

❑ A decrease in the demand for non-renewable resources, possibly by developing renewable substitutes. Some plastics, for example, could be derived from plants rather than petrochemicals.

❑ A decrease in the demand for those renewable resources that are currently being harvested at excessively high rates, so that demand could be met within sustainable yield rates. Higher value uses and higher prices for tropical hardwoods, for example, would tend to reduce demand.

Such decreases in demand could be achieved in a number of ways, including the following:

❑ An improvement in the efficiency with which forms of natural capital are converted for human use.

❑ An absolute reduction in demand. This might be achieved by developing alternative means for satisfying the relevant human needs and aspirations.

❑ The prices for some commodities could be raised until demand stabilised at levels that were within sustainable yield rates.

Managing the demand for energy

The preferred development option would be one that achieved one or more of the following:

❑ The development of means for recovering and recycling energy within the economic system.

❑ The development and implementation of renewable energy technologies.

❑ A decrease in the demand for energy, with a target of reduction to the point where demand could be met from renewable energy sources.

Such decreases in demand could be achieved in a number of ways, including the following:

❑ The development of more efficient means and technologies for achieving given ends.

❑ Improving the general efficiency of energy conversion, transmission and use.

❑ An absolute reduction in demand. This might be achieved by developing alternative means for satisfying the relevant human needs and aspirations.

❏ The prices for some energy sources could be gradually raised to encourage greater conservation and efficiency.

Environmental impact reduction

Until now, it has in some cases been necessary to prove that environmental damage was occurring before action would be taken. A transition to a more sustainable way of life will require a more precautionary approach. However, this precaution cannot be allowed to stultify development. This is because a great deal of new development will be required, as polluting production methods are replaced with cleaner technology, for example. This can create dilemmas. One way of dealing with these, as practised in parts of Germany, is to place an onus of environmental responsibility on developers. Some local authorities have one department that provides advice and support to developers, and another that ensures compliance. The German experience indicates that such departments should be kept separate, as developers will not disclose information if they feel that it might be used in a legal action against them.

Environmental impacts should be audited and assessed at both local and non-local level, over the short and long-term, in terms of both direct and indirect effects, and in terms of both the project itself and its upstream and downstream consequences. It is important to consider both the development itself and any effects that may be direct consequences of the development. New out-of-town shopping centres, for example, have consequences in terms of increased demand for transport and the associated pollution. It is important to regard this as a continuous process. The base position should be established with the first audit, while subsequent audits will monitor change. Indicators can be refined and improved throughout this process.

The preferred development option would then be that which achieved any or all of the following:

❏ eliminated wastes and approached zero emissions on either a plant or site level;
❏ reduced the toxicity of residual waste output and the volumes of waste produced to within the absorption or buffering capacity of the site;
❏ achieved these toxicity and volume reductions with a transition to cleaner technology, rather than an end-of-pipe approach;
❏ utilised optimal disposal routes for residual wastes, which achieved effective dilution factors, effective absorption capacity or effective containment as appropriate;
❏ utilised sites that were less environmentally sensitive in terms of ecological, scenic, recreational or other appropriate value;
❏ minimised biological and ecological impacts; this would be especially important when the ecological impact concerned would be critical (by being particularly extensive, for example, or by affecting the base of a food chain);

❑ had a low risk and a good worst-case profile;
❑ contributed a planning gain of environmental compensation or enhancement. Developments that entailed the loss of trees, for example, could be accompanied by compensatory planting (not necessarily on the same site) as a condition of consent. Other development losses could be appropriately compensated.
Sympathetic tree planting is a good candidate for a standard compensatory measure. This is because trees are a 'regenerative agent' in that they will absorb while growing some of the excess carbon dioxide in the atmosphere, while also encouraging wildlife, providing a source of biomass for future use, helping to displace imports of timber and so on. Of course, such planting schemes have spatial distributional consequences, and would have to be integrated into other land use demands.

The role of the individual and the community

One issue, which continues to divide the disparate environmental groups in the UK and abroad, is whether social change is brought about through change at the level of the individual or that of society, and whether the focus of action should therefore be at one level or the other. There are a range of positions in this debate.

One argument can be paraphrased as follows. The kinds of change involved are so extensive that it is hard to see how any political party could ever get into a position to carry them out. Expectations of rising standards of living, measured in terms of increased consumption, have become entrenched. Anyone who asks people to stabilise their levels of material consumption, and accept that they will not be able to consume more at the end of their lives than they could at the beginning, and that their children will not be able to consume more than they could, or who goes even further and asks people to accept a reduction in their levels of material consumption from today's levels, is unlikely to be as politically popular as someone who promises rises in standards of living. In a democracy, therefore, people will not vote for a radical programme of change unless they are somehow persuaded that it has become necessary to put their own interests second.

If attitudes to consumption have become deeply entrenched, and if people cannot be persuaded that the need for change outweighs their personal interests, then more authoritarian government might be necessary if some sort of stringent austerity measures prove to be necessary. Many people would find this an unattractive prospect. Unless there is general agreement that such a government is required, this scenario is therefore unlikely to arise though the normal democratic process. It is also uncertain that such a government could maintain governance for long enough to implement a programme. Such a government would have to be able to secure and maintain at least the minimum level of support needed to survive. This implies that the programme for change would have to be generally seen as worthy of support (such as with war-time levels of taxes and privations), at least by a sufficiently numerous or powerful section of the community.

Furthermore, simple reductions in consumption might not be enough. It may not be necessary to reduce some kinds of material consumption at all. Where reductions do prove to be necessary, it may be as important to think about the process of change and the rate and timing of change as it is to think about setting targets. If we must address the global warming problem, for example, we could easily set an arbitrary maximum emission rate for carbon. We would also have to ask a range of ancillary questions, however, as to how we should aim to achieve the reductions, what targets should be set for particular countries, areas, sectors and industries, whether the targets should be determined in terms of ecological or economic criteria, whether we should implement least-cost measures first, or examine more radical product and process redesigns, where our efforts and capital should be invested for the best prospect of success, and whether we should reconstruct our houses, industries, or modes of transport first.

The complexity of the task implies a diversity of inputs of opinion and expertise, whereas the need to maintain governance with an unpopular political programme implies a reduction in the number of routes and level of access into government. The question that arises, therefore, is whether a more authoritarian government would necessarily have the wisdom, integrity, and access to the necessary diversity of opinion and skills to devise a sophisticated and coherent public programme for change that would have the necessary effect.

This argument generally concludes that we should look less to social and political change, and more to change at the individual and family level. This kind of grass roots change involves such things as education for oneself and others, and changing attitudes, aspirations, and patterns of consumption. In terms of practical actions, one could use the car less, and the bicycle, train or bus more, compost biodegradeable waste, recycle glass, cans, paper and plastics wherever possible, insulate and draught-proof the house, turn down the thermostat, use the shower instead of the bath, and switch to buying environmentally-friendlier products. One could also become involved in helping to effect political change, lobbying and demanding action of the local MP and councillors. There are a great many potential campaign issues: local environmental standards and pollution, the provision of recycling facilities, switching resources from provision for cars to provision for bicycles and buses, creating traffic-free zones, higher insulation standards for houses and larger energy efficiency grants for low income households, the opening-up of wildlife corridors into cities and so on.

In the longer term, this kind of active involvement could also help to change attitudes and the general climate of opinion. Over a sufficient period of time, one could hope to change social norms, so that action seen as acceptable today (such as urban car use) might come to be seen as unacceptable as smoking in non-smoking public areas has started to become in the last decade.

An alternative argument is founded on the concern that the pace of global environmental change is now so fast, and the impact so extensive, that the kind of bottom-up approach based on individual change and action will simply be too slow. Fundamental change in attitudes, such as a weakening of the link between status and success and patterns of material consumption, may require several generations to achieve. However, rates of population growth, global warming, and so on, indicate that widespread ecological disaster of some kind could be with us in rather less time.

Political change, therefore, will be essential if we are to be able to change the future course of society onto a sustainable development pathway within the time available. Tough decisions will probably have to be taken, resources may have to be switched between sectors, or from consumption into investment, imports and exports controlled and so on.

This would affect the domestic distribution of wealth. Such changes could also affect the international flows of resources, and hence the international flows of capital, which in turn could affect the distribution of power and wealth between nations. The fact that some existing interests will try to resist any encroachment on their profits, power or privileges that would result from such changes indicates that it would be necessary to have a political decision-making system that is sufficiently strong and robust to be able to resist the pressures that such interests may apply.

The authors incline to a synthesis of these arguments. We believe that action has to be taken on different levels, and that changes made in only one area of life would probably not be sufficient. Consider waste minimisation and recycling, for example. At the level of the individual or the family, people could change their purchasing patterns in order to cut down on the amount of waste that they generate. They might cut down on inessential purchases, buy fewer but longer-life goods, buy items loose rather than in pre-packs and so on. Even long-life items fail eventually, however, and some packaging is essential (for perishable or sterile products, for example), so even the most committed reformer will continue to generate some waste. Some waste can perhaps be re-used by the consumer for some other purpose, but much will still have to go for further processing before it could be re-used. It makes little sense for individuals to try to do this on their own. At this stage, therefore, a degree of social organisation is required. If wastes are to be recycled efficiently, they must be collected, sorted, screened and processed in relatively large quantities. The necessary infrastructure of vehicles, plant, collection points and so on would have to be put in place. Similarly, it is hard to argue that people should use their cars less if their local railway line has been closed and their bus service is poor. The necessary public transport infrastructure would have to be in place, so that people had a viable and at least equally attractive alternative to the use of a private vehicle (it does not matter, in this context, whether the new waste recycling plant or the public transport system is built or owned by the public or private sectors, although it is unlikely that the private sector would collect wastes or operate transport links on any but the most profitable routes without some degree of government guidance and funding).

Thus a combination of individual commitment and social action will be required to effect change.

The most fundamental point, however, is that it is vital to ensure that people are not penalised for good behaviour and rewarded for bad. In a situation in which environmentally-responsible citizens are punished for their attempt to use the public transport system with a slow and inferior service, while the environmentally-careless citizens are subsidised with cheap roads and petrol to use their cars, it is hardly surprising that the wishes of the individual and the needs of society and the environment are at odds.

Of course, people should be made aware of the issues, and encouraged to behave in an environmentally-responsible way. It might be unwise to assume, however, that everyone will always behave in an environmentally responsible way. The best solution, therefore, would be to make it easier and cheaper to to do the right thing (with good recycling and public transport facilities, for example), and harder and more expensive to do the wrong thing (with refuse collection charges and road pricing, for example), thereby ensuring that the schedule of costs and benefits for the individual were aligned with those of society and the environment. Some might argue that good behaviour should not be subsidised, but few could argue against the case for identifying and stripping out any subsidies for bad behaviour. This would make individuals more likely to behave in an environmentally responsible way simply by responding to market signals.

Individual and social values

A more fundamental reason why change is needed at both individual and societal levels is that the change at these levels should be mutually reinforcing, and should therefore, ideally, be as consistent as possible in order to convey a coherent message. It is harder to bring children up with a sense of ethics and values, for example, if society then rewards aggressive, selfish and greedy behaviour.

In any social system behaviour is shaped, to an extent, by a system of positive and negative sanctions. Behaviour that is held to be desirable is valued, recognised or rewarded. Behaviour that is held to be undesirable is punished in some way, such as by killing, imprisonment, or exclusion. As children grow up in a given social system, they internalise these values, so that they come to judge and regulate their own behaviour by these standards. Of course, such behavioural codes do themselves change over time, partly in response to changing needs and circumstances. It is also possible for individuals to refuse to conform to any given code. As we are a social species, we generally find it hard to exist without group recognition and support. When people reject a given norm, therefore, they usually do it by adopting the behavioural norms of a deviant sub-group with similar views within the larger community.

Pre-industrial societies can operate without a formal system of policing, using only informal checks and controls, because in such societies behavioural codes and customs tend to be relatively strong, with a high degree of consensus. In a more differentiated society, there may no longer be just one agreed behavioural code. The process of industrialisation itself transforms many of the relationships between individuals and groups within society. Existing customs generally do not continue to serve as adequate guides to behaviour throughout such social transformations.

A more long-term solution to many of the global environmental problems might well entail a degree of change in both individual attitudes and in the social mechanisms used to shape behaviour, such as the schedule of rewards and punishments for particular kinds of behaviour. This is not an argument for 'social engineering', as that implies imposing a uniformity of view, whereas, as has been argued elsewhere, a transition to a more sustainable way of life implies a multiplicity of views.

It is, however, an argument for developing a greater emphasis on civics, politics, economics and ethics in our educational systems, in order to foster a greater understanding of the relationship and tensions between the needs of the individual and those of society.

The role of the educator

There are moral questions that pervade much of the sustainability debate. They are especially obvious in intergenerational issues. Why, for example, should we assume a responsibility to leave the world in any particular condition for future generations?

An actual change to a more sustainable way of life is likely to encounter a number of such issues and dilemmas, and may therefore require a degree of change in understanding and attitudes. Education is one of the ways in which moral values and positions are developed in society. A transition to sustainability may therefore require some change in current educational programmes.

The process of education is, of course, not confined to schools and universities. Education and training at all levels will assume, inculcate, or otherwise develop positions from which value judgements are made. A different approach would be required in the various sectors.

❑ Education in environmental issues, civics, politics, economics and ethics in our schools would help to create an educated citizenry, capable of making the decisions that future society will face.
❑ Interdisciplinary teaching and research, the study of system dynamics, ethics, and the philosophy, methods and principles of science would help to bridge some of the disciplinary divides in our universities and colleges, and to bring that wealth of expertise and knowledge to bear on the great social and environmental issues of the day.
❑ A more applied approach might be appropriate in the business and industrial sectors. One essential task would be to develop a wider concept of social and environmental responsibility. There are a number of techniques that would be useful in this regard, such as the Japanese *ringi* system and strategic choice planning methods, perhaps in conjunction with personal development training for staff with a focus on the development of a social and environmental awareness.

The agenda for further research

The authors believe that research that addressed any of the following issues would be helpful.

Scientific and technical questions

We need to know as much as possible about the key environmental problems, and about any other environmental issues that we have, as yet, failed to identify. We need to know how serious the problems are, what the associated risks are, and about the causes. It will be necessary to clarify degrees of uncertainty, and to try to identify the critical areas of ignorance. It is especially important to try to identify areas of possible indeterminate behaviour in the environmental systems on which we depend.

We also need information about trends, to know how quickly situations are changing, whether the rate of change is itself accelerating or decelerating, and to have estimates of the time remaining until the point at which such environmental change might impact on society.

One of the most fundamental objects of such research questions is to try to improve our understanding of complex adaptive systems in general and of the global ecology in particular to the point where we can define the degrees of freedom or 'environmental space' available to the human species, that is, the area within which we can operate (in terms of extracting flows of resources and utilising pollution absorption capacity) without imposing undue stress on the system as a whole.

There are a number of related technological questions. Part of a necessary response to environmental pressure, for example, might be to develop or disseminate new technologies as substitutes for existing technologies. Given the extraneous factors that, in practice, usually determine whether a particular technology is developed, this may be as much a political and economic issue as a technical one.

Political and economic questions

We need more research on the implications of the changes that might be necessary in order to achieve a sustainable development pathway. Research will be needed to identify how and where such changes would impact on existing structures, to identify groups that would gain by such changes and groups that would lose, to ascertain whether some people would be more significantly affected than others, and the extent to which there would be support for any programme of help or compensation for such people, and to estimate the consequences for the domestic and international distribution of wealth, power and resources.

Research is also needed on the ways to encourage change. It would be useful to know, for example, the combination of penalties and incentives that would best encourage the productive sectors of the economy to innovate and develop in the direction of greater energy and resource-use efficiency and cleaner technology, whether a regulatory or more market-oriented approach would be more effective, and what kind of relationship between the regulator and the regulated would be most effective in achieving the desired outcome.

This in turn will require a better understanding of the processes of decision-making on environmental issues in industry. It would be useful to know, for example, how the issues and pressures were perceived, which sources of pressure were most significant, and what combination of regulatory pressures, economic pressures, and factors in both the internal and the external cultural environment of the firm were effective in forcing or enabling innovation and development in this particular direction.

There are deeper issues as to where and to what extent we can depend on scarcity to push up prices and thereby automatically encourage the development of such new resources and technologies, that is, the extent to which Adam Smith's 'invisible hand' will guide the markets towards greater energy and resource-use efficiency, and where and to what extent this process will be deficient and where explicit proactive policies will be needed to encourage, for example, the development of new resources and technologies.

There are other fundamental questions as to how to balance economic and environmental costs, the extent to which environmental performance should be improved, where it might be necessary to take more radical action than would be justified by a strict calculation of costs and benefits and, of course, the basis on which such costs and benefits should be calculated.

Ethical and philosophical questions

We need more thinking and general debate on the deep ethical dilemmas at the root of the problems.

We need to ask, for example, to what extent the current generation should assume responsibility for the global ecological crisis, and why they should behave differently from previous generations in this regard.

There are even deeper questions as to the purpose and meaning of human existence, the nature of happiness and so on. There is a question, for example, as to whether we need to or should consume all the resources available to us, with or without allowing for the needs of future generations. We must consider the discrepancy between the assumption made by many neoclassical economists, that happiness can be equated with consumption, with the belief held in many of the world's religious and philosophical traditions, that happiness and fulfillment cannot be achieved by maximising material consumption. We must further consider the discrepancy between the neoclassical position and the psychological and sociological research that indicates that a sense of self-worth, recognition and relative position within society are often of more importance than absolute levels of material consumption. It may be necessary to develop, at the end of this process, a new philosophical basis for economics.

There are also questions as to values. It might be argued, for example, that a future generation that inherited a degraded and depleted environment, but also inherited the benefits and wealth that were created by exploiting that environment would have been adequately compensated, especially given that they would have no first-hand knowledge of what had been lost. Others might argue that the current generation have no moral right to make that decision on behalf of the future generation, or that the compensation would not be adequate, or that humans have no moral right to exterminate other species.

It is also possible to argue that, if the current generation have no moral right to impose the consequences of an unsustainable way of life on a future generation, that a particular group within the current generation have no moral right to impose the requirements of a sustainable development pathway on the rest of the current generation. It is necessary to consider, therefore, how to weigh up the right of people to behave as they wish against the well-being of the environment, or of future generations.

These arguments should be developed; they form a central part of the sustainability debate.

Further Reading

Robert Costanza and Lisa Wainger (eds) *Ecological Economics: the Science and Management of Sustainability* Columbia University Press, New York, 1991

Robert Costanza et al *The ecological economics of sustainability: making local short-term goals consistent with global and long-term goals* World Bank Environment Department: Environment Working Paper No 32, 1990

Herman E Daly and John B Cobb Jr *For the Common Good: Redirecting the Economy towards Community, the Environment and a Sustainable Future* Green Print, London, 1990

Paul Ekins (ed) *The Living Economy: a New Economics in the Making* Routledge and Kegan Paul, London, 1986

Murray Gell-Mann Part IV of *The Quark and the Jaguar: Adventures in the Simple and the Complex* Little, Brown, 1994

Nick Hanley and Clive L Spash *Cost–Benefit Analysis and the Environment* Edward Elgar, 1993

Will Hutton *The State We're In* Jonathan Cape, London, 1995

Michael Jacobs *The Green Economy* Pluto, London, 1991

Roger Lewin *Complexity: Life at the Edge of Chaos* J M Dent Ltd, London, 1993

D Meadows, D Meadows and J Randers *Beyond the Limits* Earthscan, London, 1992

P Ormerod *The Death of Economics* Faber and Faber, London, 1994

D Pearce, A Markandya and E Barbier *Blueprint for a Green Economy* Earthscan, London, 1989

John Peet *Energy and the Ecological Economics of Sustainability* Island Press, Washington DC, 1992

John Pezzey *Economic analysis of sustainable growth and sustainable development* World Bank Environment Department: Working Paper No 15, 1989

Karl Sigmund *Games of Life: Explorations in Ecology, Evolution and Behaviour* Oxford University Press, 1993

M M Waldrop *Complexity: The Emerging Science at the Edge of Order and Chaos* Viking, London, 1993

References

1   Jonathan Silvertown 'Earth as an environment for life', in Jonathan Silvertown and Philip Sarre, editors *Environment and Society* Hodder and Stoughton (London) 1990

2   R I Moore, editor *Philip's Atlas of World History* Octopus Illustrated Publishing (London) 1992

3   Lewis Mumford *The City in History* Pelican (Harmondsworth) 1961

4   Clive Ponting *A Green History of the World* Sinclair-Stevenson (London) 1991

5   Stephan Jay Gould *The Flamingo's Smile: Reflections in Natural History* WW Norton (New York) 1985

6   Paul Wignall 'The day the world nearly died' *New Scientist*, 133 (1805): 51–55 (London) 25 January 1992

7   P B Medawar and J S Medawar *The Life Science* Wildwood House (London) 1977

8   Gerald M Weinberg *An Introduction to General Systems Thinking* John Wiley and Sons Inc (New York) 1975

9   L von Bertalanffy 'The organism considered as a physical system' in von Bertalanffy *General System Theory: foundations, development, applications* Braziller (New York) 1969

10 H Simon 'The architecture of complexity' in Simon *The Sciences of the Artificial* MIT Press (Cambridge, Mass) 1969

11 Jared Diamond *The Rise and Fall of the Third Chimpanzee* Vintage (London) 1991

12 J A S Kelso and K G Holt 'Exploring a vibratory systems account of human movement production', *Journal of Neurophysiology*, 43:1183–1196, The American Physiological Society (Bethesda) 1980

13 J A S Kelso, K G Holt, P Rubin, and P N Kugler 'Patterns of human interlimb coordination emerge from the properties of non-linear limit cycle oscillatory processes: theory and data' *Journal of Motor Behaviour*, (13(4):226–261 Heldref Publications (Washington) 1981

14 N Jordan *Themes in Speculative Psychology* Tavistock (London) 1968

15 Peter Checkland *Systems Thinking, Systems Practice* John Wiley and Sons (Chichester) 1981

16 Paul Ormerod *The Death of Economics* Faber and Faber (London) 1994

17 N Duncan 'Why can't we predict?' *New Scientist*, 136(1841):47, (London) 1992

18 D H Meadows, D L Meadows, and J Randers *Beyond the Limits* Earthscan (London) 1992

19 James Gleik *Chaos: Making a New Science* Viking (New York) 1987

20 Richard Dawkins *The Extended Phenotype* Oxford University Press (Oxford) 1982

21 Richard Dawkins *The Blind Watchmaker* Penguin (Harmondsworth) 1988

22 John H Holland *Adaptation in Natural and Artifical Systems* University of Michigan Press (Ann Arbor) 1975

23 Shelagh Ross 'Atmospheres and climatic change' in Paul M Smith and Kiki Warr, editors, *Global Environmental Issues* Hodder and Stoughton (London) 1991

24 James E Lovelock *Gaia: a new look at life on Earth* Oxford University Press (Oxford) 1979

25 Roger Lewin 'Living in a bubble' *New Scientist*, 134(1815):12–13, (London) 4 April 1992

26 Gregory Bateson *Steps to an Ecology of Mind* Paladin (London) 1973

27 John D Barrow and Frank J Tipler *The Anthropic Cosmological Principle* Oxford University Press (Oxford) 1986

28 Ian Stewart, *Does God Play Dice?* Penguin Books (Harmondsworth) 1990

29 R Ambroggi 'The water of our planet' Technical report, Centre d'Etudes Pratiques de la Negociation Internationale (Geneva) 1985

30 Sandy Smith and Owen Greene 'The Oceans' in Paul M Smith and Kiki Warr, editors, *Global Environmental Issues* Hodder and Stoughton (London) 1991

31 David Olivier, David Elliot, and Alan Reddish 'Sustainable energy futures' in John Blunden and Alan Reddish, editors, *Energy, Resources and Environment* Hodder and Stoughton (London) 1991

32 E Linden 'The world's water' *Time International* (New York) 5 November 1990

33 J T Houghton, J Jenkins, and J Ephraums *Climate Change: the IPCC scientific assessment* Cambridge University Press (Cambridge) 1990

34 P Brookes and D Jenkinson 'The enhancer lies in the soil' *The Guardian* (London) 3 April 1992

35 G Walker 'Diluted ocean threatens Western Europe's weather' in *New Scientist*, 148(2003) (London) 11 November 1995

36 J Gribbin 'Global warming cuts no ice in Greenland' in *New Scientist*, 148(2003) (London) 11 November 1995

37 World Resources Institute *World Resources 1986*, WRI (Washington) 1986

38 Professor Brian Wynne, Seminar at the Research Centre for Social Science, University of Edinburgh , 28 February 1994

39 Kim Van Skoy and Kenneth Coale 'Pumping iron in the Pacific' *New Scientist*, 144(1954) (London) 3 December 1994

40 Fred Pearce 'Iron soup feeds algal appetite for carbon dioxide' *New Scientist*, 147(1984) (London) 1 July 1995

41 Kiki Warr 'The ozone layer' in Paul M Smith and Kiki Warr, editors, *Global Environmental Issues* Hodder and Stoughton (London) 1991

42 Fred Pearce 'Clean air will expose Europe to global warming' *New Scientist* 133(1805):17 (London) 25 January 1992

43 HMSO *Sustainable Development, the UK Strategy* HMSO (London) 1994

44 Jonathan Silvertown 'Inhabitants of the biosphere' in Jonathan Silvertown and Philip Sarre, editors, *Environment and Society* Hodder and Stoughton (London) 1990

45 Owen Greene 'Tackling global warming' in Paul M Smith and Kiki Warr, editors, *Global Environmental Issues* Hodder and Stoughton (London) 1991

46 Stephen Jay Gould *Bully for Brontosaurus: Reflections in Natural History* Hutchinson Radius (London) 1991

47 Roger Lewin 'Life and death in a digital world' *New Scientist* 133(1809):36–39 (London) 22 February 1992

48 Roger Lewin 'How to destroy the doomsday asteroid' *New Scientist* 134(1824):12–13 (London) 6 June 1992

49 Edward O Wilson 'Threats to biodiversity' *Scientific American* 261(3):60–66 (New York) September 1989

50 Wolfgang Lutz and Christopher Prinz 'New world population scenarios' Technical report, International Institute for Applied Systems Analysis (New York) Autumn 1994

51 J Caldwell *Theory of Fertility Decline* Academic Press (London) 1982

52 David Grigg 'World agriculture: productivity and sustainability' in Philip Sarre, editor, *Environment, Population and Development* Hodder and Stoughton (London) 1991

53 BBC2: *The Natural World* 'Prisoners of the sun' 19 January 1992

54 Herman E Daly 'Toward some operational principles of sustainable development' *Ecological Economics*, 2:1–6 (Amsterdam) 1990

55 Paul R Ehrlich and Anne H Ehrlich *Population Resources Environment: Issues in Human Ecology* W H Freeman and Company (San Francisco) 1970

56 Herman E Daly and John B Cobb Jr *For the Common Good: Redirecting the Economy Towards Community, the Environment and a Sustainable Future* GreenPrint (London) 1990

57 John Gever, Robert Kaufman, David Skole, and Charles Vorosmarty *Beyond Oil: The Threat to Food and Fuel in the Coming Decades* Ballinger (Cambridge, Mass) 1987

58 Talbot Page *Conservation and Economic Efficiency* John Hopkins University Press (Baltimore) 1977

59 Tablot Page 'International equity and the social rate of discount' in V K Smith, editor, *Environmental Resources and Applied Welfare Economics* Resources for the Future (Washington DC) 1988

60 Charles W Howe *Natural Resource Economics: Issues, Analysis and Policy* John Wiley and Sons (New York) 1979

61 John C Bergstrom 'Concepts and measures of the economic value of environmental quality: a review' *Journal of Environmental Management* 31:215–228 (London) 1990

62 D S Brookshire and D L Coursey 'Measuring the value of a public good: an empirical comparison of elicitation procedures' *American Economic review*, 77:554–565 The American Economic Association (Princeton) 1987

63 R G Cummings, D S Brookshire, and W D Schulze *Valuing Environmental*

Goods: an Assessment of the Contingent Valuation Method Rowman and Littlefield (Maryland) 1986

64 Clive L Spash *Greenhouse Economics: Value and Ethics* Routledge (London) forthcoming

65 Jack L Knetsch 'Issues in environmental valuation' *Green Values* Scottish Environmental Economics Discussion Group (University of Stirling) Spring 1993

66 Paul A Samuelson *Economics* McGraw Hill Kogakusha (International Edition) (Tokyo) 1980

67 A C Pigou *The Economics of Welfare* Macmillan (London) 1920

68 HMSO *The Abolition of Domestic Rates (Scotland) Act* HMSO (London) 1987

69 HMSO *The Local Government Finance Act* HMSO (London) 1988

70 Wilfred Beckerman 'Pricing for pollution' Technical report, Institute of Economic Affairs (London) 1990

71 Salah El Serafy 'The proper calculation of income from depletable natural resources' in Ernst Lutz and Salah El Serafy, editors, *Environmental and Resource Accounting and their Relevance to the Measurement of Sustainable Income* World Bank (Washington DC) 1988

72 Rick Worrell 'Trees and the treasury: Valuing forests for society' Technical report, World Wide Fund for Nature (Godalming) 1991

73 John Pezzey 'Economic analysis of sustainable growth and sustainable development' Technical report, World Bank Environment Department Working Paper No 15 World Bank (Washington DC) March 1989

74 Will Hutton 'Labour's square mile target' *The Guardian* (London) 23 March 1992

75 John Kenneth Galbraith 'Money: whence it came, where it went' Viking (New York) 1975

76 George Edwards 'British banks do not give credit where it is due' *The Guardian* (London) 10 February 1992

77 Reviewed in *Options*, The Journal of the International Institute for Applied Systems Analysis, IIASA, Austria, Summer 1995

78 S Day 'Invasion of the Shapechangers' *New Scientist*, 148(2001) (London) 28 October 1995

79 Roger Lewin 'Making maths make money' *New Scientist*, 134(1816):31–34, 11 April 1992

80 Brian Arthur 'Positive feedbacks in the economy' *Scientific American* pp 92–99, (New York) February 1990

81 Alun Lewis and Tessa Lecomber (for Channel 4) 'Antichaos and the science of complexity', 1992

82 Paul Wallich and Elizabeth Corcoran 'The analytical economist' *Scientific American* 261(3):58 September, 1989

83 Paul M Smith 'Sustainable development and equity' in Paul M Smith and Kiki Warr, editors, *Global Environmental Issues* Hodder and Stoughton (London) 1991

84 J K Galbraith *Money: Whence it came, where it went* Penguin (Harmondsworth) 1975

85 Paul M Smith 'Global development issues' in Paul M Smith and Kiki Warr, editors, *Global Environmental Issues* Hodder and Stoughton (London) 1991

86 Ruben A Mnatsakanian *Environmental Legacy of the Former Soviet Republics* Centre for Human Ecology, University of Edinburgh, 1992

87 World Commission on Environment and Development *Our Common Future* (New York) 1987

88 Will Hutton 'The Student becomes the master' *The Guardian* (London) 25 November 1991

89 Monir Tayeb *Organizations and National Culture: a Comparative Analysis* Sage (London) 1988

90 Will Hutton and Ruth Kelly 'Lending: play it smart, but play it long' *The Guardian* (London) 13 May 1992

91 The Economist 'Building blocks or stumbling blocks?' *The Economist*, 325(7783):85 (London) 31 October 1992

92 Radio 4: Analysis, 4 June 1992

93 Will Hutton 'A troubled new world' *The Guardian* (London) 3 February 1992

94 R H Gray *The Greening of Accountancy: the Profession After Pearce* (ACCA Certified Research Report 17) Certified Accountants Publications Ltd (London) 1990

95 J von Neumann and O Morgenstern *Theory of Games and Economic Behaviour* Princeton University Press (Princeton) 1944

96 Douglas R Hofstader *Metamagical Themas: Questing for the Essence of Mind and Pattern* Viking (Harmondsworth) 1985

97 Robert M Solow 'The economics of resources or the resources of economics' *American Economic Review*, 64(2):1–14 The American Economics Association (Princeton) 1974

98 Stephen A Marglin 'The social rate of discount and the optimal rate of investment' *Quarterly Journal of Economics.* 77:95–111 MIT Press (Cambridge, Mass) February 1963

99 The Economist *The Economist Book of Vital World Statistics* The Economist Books (London) 1990

100 Gunnar Myrdal 'Institutional economics' *Journal of Economic Issues*, Association of Evolutionary Economics (University of Nebraska) 12:771–783 December 1978

101 R Mestel 'Let mind talk unto body' *New Scientist*, 143(1935):26–31 (London) 23 July 1994

102 A Glyn 'Decade that made the rich more equal' *The Guardian* (London) 8 June 1992

103 J R Cambell *Robert Burns – The Democrat* Hampden Advertising Ltd (Glasgow) 1945

104 Clive L Spash and Anthony M H Clayton 'Strategies for the maintenance of natural capital: sustainability, markets and ethics' Ecological Economics Discussion Paper (University of Stirling) 1995

105 William K Kapp *The Social Cost of Private Enterprise* Schoken (New York) 1950

106 Laurence A Tribe 'Policy science: Analysis or ideology' *Philosophy and Public Affairs*, 2:66–110 Princeton University Press (Princeton) fall 1972

107 Ezra J Mishan 'How valid are the economic evaluations of allocative changes?' *Journal of Economic Issues*, 14:143–161 (Nebraska) March 1980

108 A D Hall *A Methodology for Systems Engineering* Van Nostrand (Princeton, NJ)

1962

109 A D Hall 'Three dimensional morphology of systems engineering' in F Rapp, editor *Contributions to a philosophy of technology* Reidel (Dordrecht) 1969

110 R de Neufville and J H Stafford *Systems Analysis for Engineers and Managers* McGraw-Hill (New York) 1971

111 R N McKean *Efficiency in Government through Systems Analysis* Wiley (New York) 1958

112 R K Merton 'Foreword' in J Ellul (author) *The Technological Society* Jonathan Cape (London) 1965

113 R K Merton 'Manifest and latent functions' in N J Demerath and R A Peterson, editors *System, Change and Conflict* Free Press (New York) 1967

114 Peter Söderbaum 'Environmental management: A non-traditional approach' *Journal of Economic Issues*, 21(1) (Nebraska) March 1987

115 M Allen 'Ecosystems for industry' *New Scientist* 141(1911):21–22, (London) 5 February 1994

Index